Succeed

Eureka Math®

Grade 2
Modules 1–3

Published by Great Minds®.

Printed in the U.S.A.
This book may be purchased from the publisher at eureka-math.org.
BAB 10 9 8 7 6 5

ISBN 978-1-64054-084-2

G2-M1-M3-S-06.2018

Learn • Practice • Succeed

Eureka Math® student materials for *A Story of Units®* (K–5) are available in the *Learn, Practice, Succeed* trio. This series supports differentiation and remediation while keeping student materials organized and accessible. Educators will find that the *Learn, Practice,* and *Succeed* series also offers coherent—and therefore, more effective—resources for Response to Intervention (RTI), extra practice, and summer learning.

Learn

Eureka Math Learn serves as a student's in-class companion where they show their thinking, share what they know, and watch their knowledge build every day. *Learn* assembles the daily classwork—Application Problems, Exit Tickets, Problem Sets, templates—in an easily stored and navigated volume.

Practice

Each *Eureka Math* lesson begins with a series of energetic, joyous fluency activities, including those found in *Eureka Math Practice.* Students who are fluent in their math facts can master more material more deeply. With *Practice,* students build competence in newly acquired skills and reinforce previous learning in preparation for the next lesson.

Together, *Learn* and *Practice* provide all the print materials students will use for their core math instruction.

Succeed

Eureka Math Succeed enables students to work individually toward mastery. These additional problem sets align lesson by lesson with classroom instruction, making them ideal for use as homework or extra practice. Each problem set is accompanied by a Homework Helper, a set of worked examples that illustrate how to solve similar problems.

Teachers and tutors can use *Succeed* books from prior grade levels as curriculum-consistent tools for filling gaps in foundational knowledge. Students will thrive and progress more quickly as familiar models facilitate connections to their current grade-level content.

Students, families, and educators:

Thank you for being part of the *Eureka Math®* community, where we celebrate the joy, wonder, and thrill of mathematics.

Nothing beats the satisfaction of success—the more competent students become, the greater their motivation and engagement. The *Eureka Math Succeed* book provides the guidance and extra practice students need to shore up foundational knowledge and build mastery with new material.

What is in the Succeed *book?*

Eureka Math Succeed books deliver supported practice sets that parallel the lessons of *A Story of Units®*. Each *Succeed* lesson begins with a set of worked examples, called *Homework Helpers*, that illustrate the modeling and reasoning the curriculum uses to build understanding. Next, students receive scaffolded practice through a series of problems carefully sequenced to begin from a place of confidence and add incremental complexity.

How should Succeed *be used?*

The collection of *Succeed* books can be used as differentiated instruction, practice, homework, or intervention. When coupled with *Affirm®*, *Eureka Math*'s digital assessment system, *Succeed* lessons enable educators to give targeted practice and to assess student progress. *Succeed*'s perfect alignment with the mathematical models and language used across *A Story of Units* ensures that students feel the connections and relevance to their daily instruction, whether they are working on foundational skills or getting extra practice on the current topic.

Where can I learn more about Eureka Math *resources?*

The Great Minds® team is committed to supporting students, families, and educators with an ever-growing library of resources, available at eureka-math.org. The website also offers inspiring stories of success in the *Eureka Math* community. Share your insights and accomplishments with fellow users by becoming a *Eureka Math* Champion.

Best wishes for a year filled with Eureka moments!

Jill Diniz
Director of Mathematics
Great Minds

Contents

Module 1: Sums and Differences to 100

Module 2: Addition and Subtraction of Length Units

Module 3: Place Value, Counting, and Comparison of Numbers to 1,000

Grade 2
Module 1

Fluency Practice

Making ten and adding to ten is foundational to future Grade 2 strategies. Students use a number bond to show the part–whole relationship with numbers.

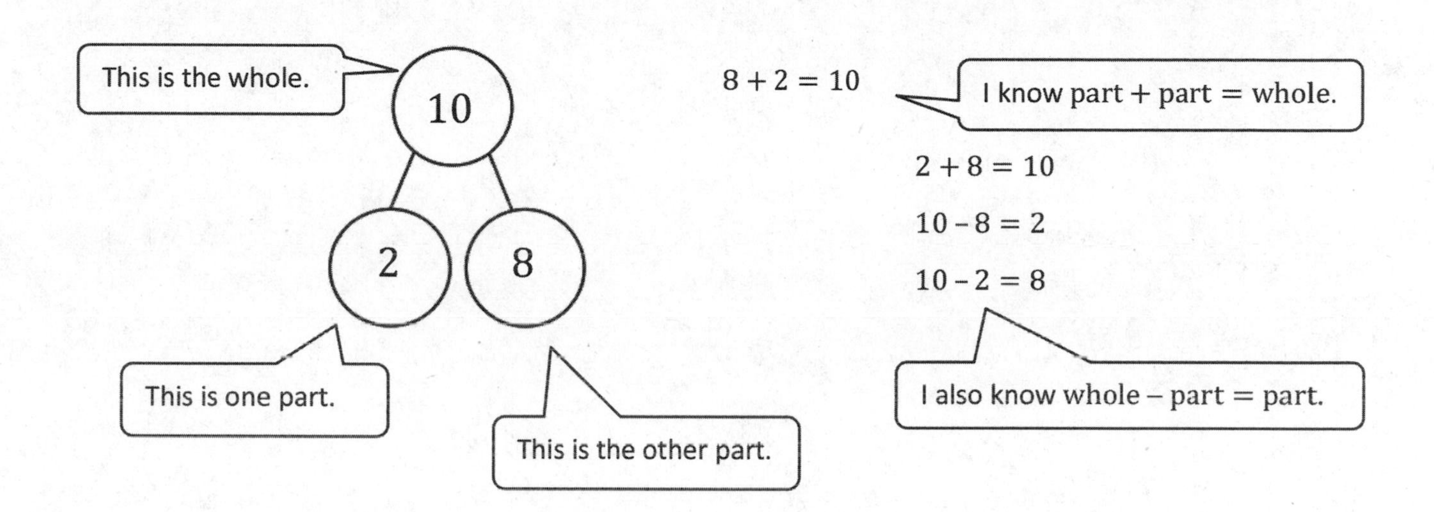

$10 = 7 + \mathbf{3}$

> I need to be careful when looking at the signs.
> This says 10 *equals* 7 + ___, not 10 *plus* 7 = ___.
> That means 10 is the same as 7 + 3.

Name ______________________________ Date ______________

1. Add or subtract. Complete the number bond for each set.

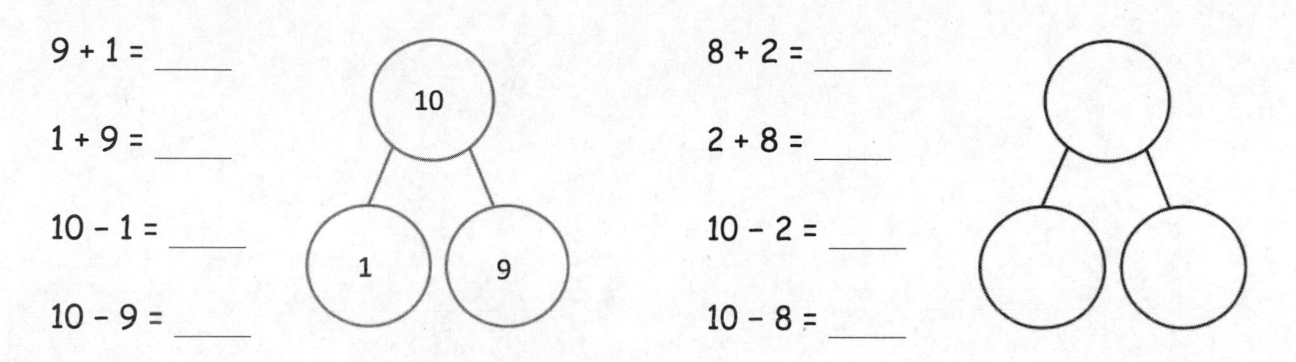

9 + 1 = ____

1 + 9 = ____

10 – 1 = ____

10 – 9 = ____

8 + 2 = ____

2 + 8 = ____

10 – 2 = ____

10 – 8 = ____

2. Solve. Draw a number bond for each set.

6 + 4 = ____

4 + 6 = ____

10 – 4 = ____

10 – 6 = ____

3 + 7 = ____

7 + 3 = ____

10 – 7 = ____

10 – 3 = ____

3. Solve.

10 = 7 + ____

10 = 3 + ____

10 = 5 + ____

10 = 2 + ____

10 = ____ + 8

10 = ____ + 4

10 = ____ + 6

10 = ____ + 1

Fluency Practice

Making the next ten and adding to a multiple of ten is foundational to future Grade 2 strategies. Students continue to use a number bond to show the part-whole relationship with numbers.

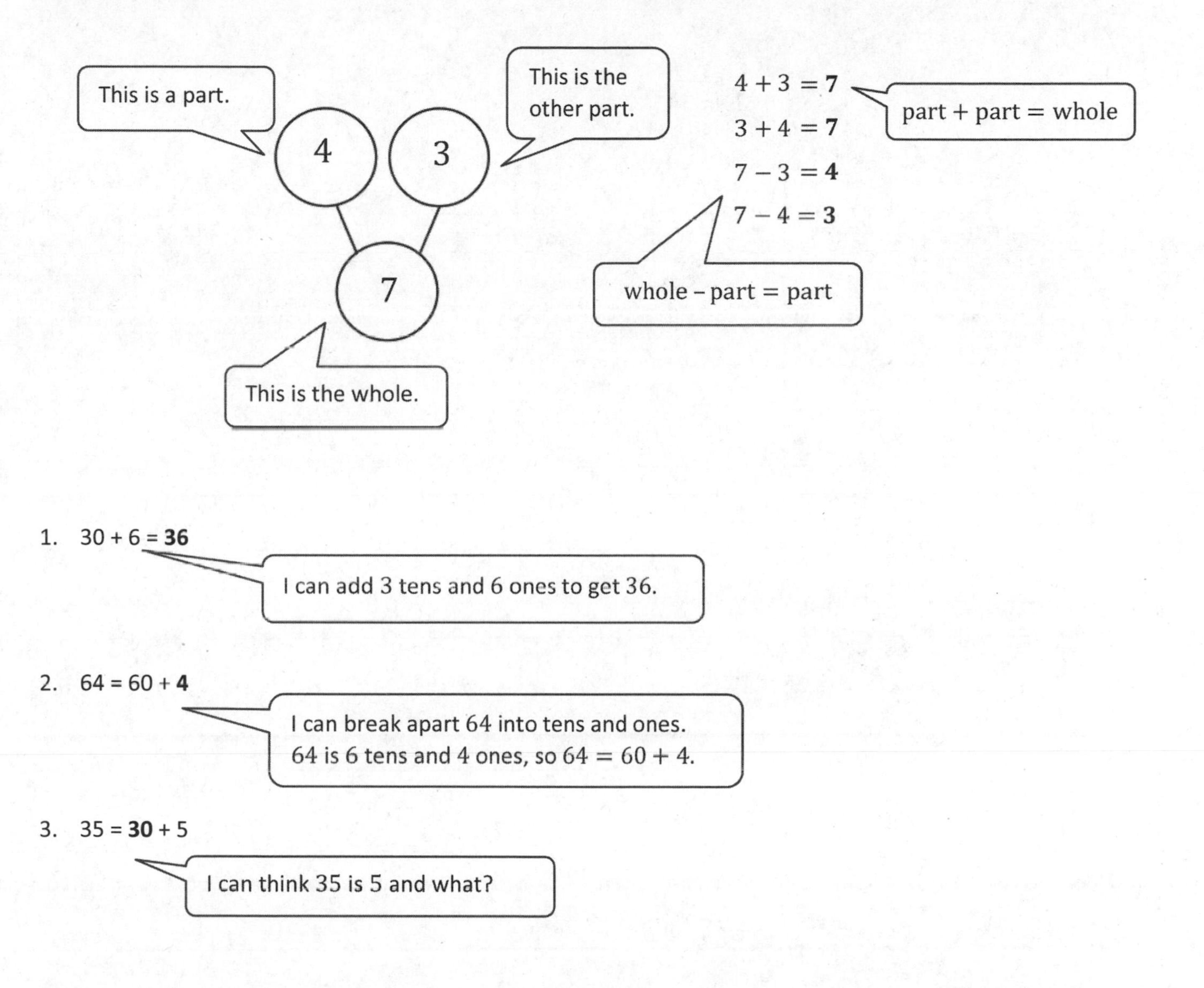

1. 30 + 6 = **36**

   I can add 3 tens and 6 ones to get 36.

2. 64 = 60 + **4**

   I can break apart 64 into tens and ones. 64 is 6 tens and 4 ones, so $64 = 60 + 4$.

3. 35 = **30** + 5

   I can think 35 is 5 and what?

Add and Subtract Like Units, Ones, To Solve Problems Within 100

1. $20 + 7 = \mathbf{27}$
2. $20 + 70 = \mathbf{90}$
3. $62 + 3 = \mathbf{65}$
4. $62 + 30 = \mathbf{92}$

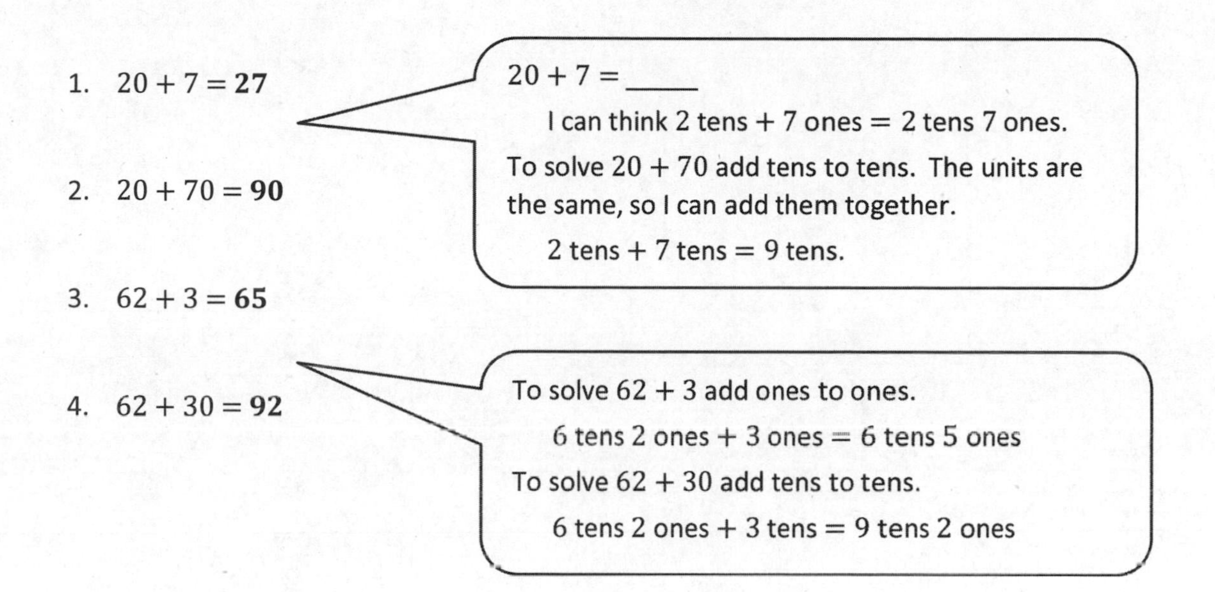

5. Complete each blank in the table below.

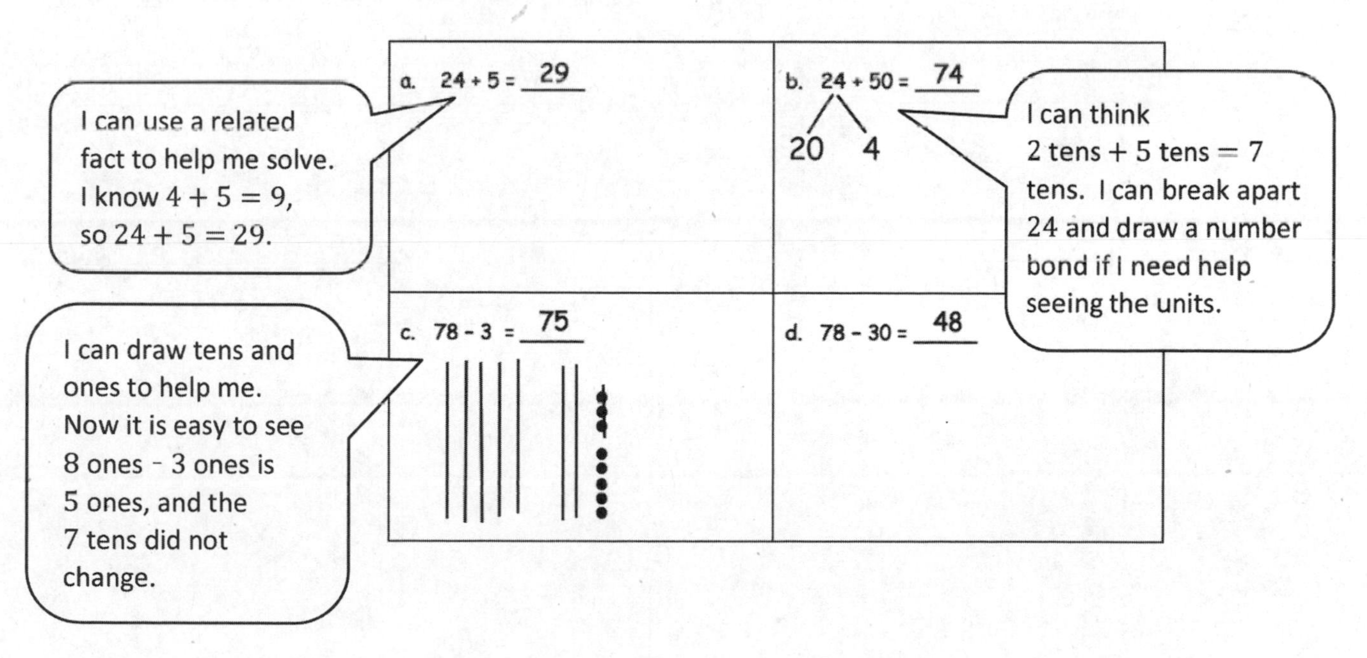

Name ____________________ Date ____________

1. Solve.

a. 20 + 7 = ____

20 + 70 = ____

62 + 3 = ____

62 + 30 = ____

b. 80 – 20 = ____

85 – 2 = ____

85 – 20 = ____

86 – 20 = ____

c. 30 + 40 = ____

31 + 40 = ____

35 + 4 = ____

45 + 30 = ____

d. 70 – 30 = ____

75 – 30 = ____

78 – 3 = ____

75 – 40 = ____

2. Solve.

a. 42 + 7 = ___	b. 24 + 70 = ___
c. 49 – 2 = ___	d. 98 – 20 = ___

3. Solve.

a. 16 + 3 = _____ 13 + 6 = _____	b. 37 – 3 = _____ 37 – 4 = _____
c. 26 + 70 = _____ 76 + 20 = _____	d. 97 – 50 = _____ 97 – 40 = _____

Making Ten from an Addend of 9, 8, or 7

1. 9 + 3 = **12**

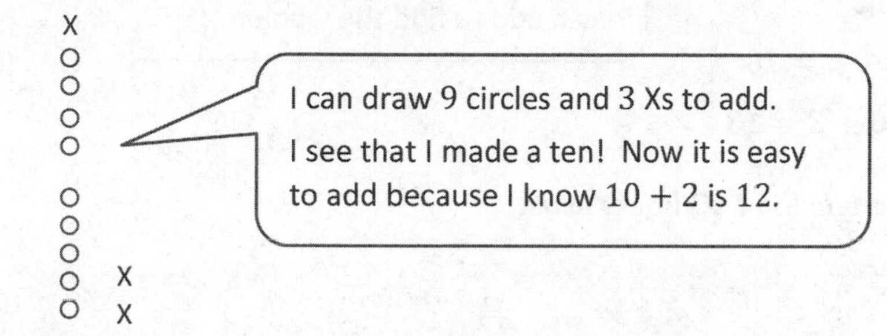

2. 8 + 7 = **15**

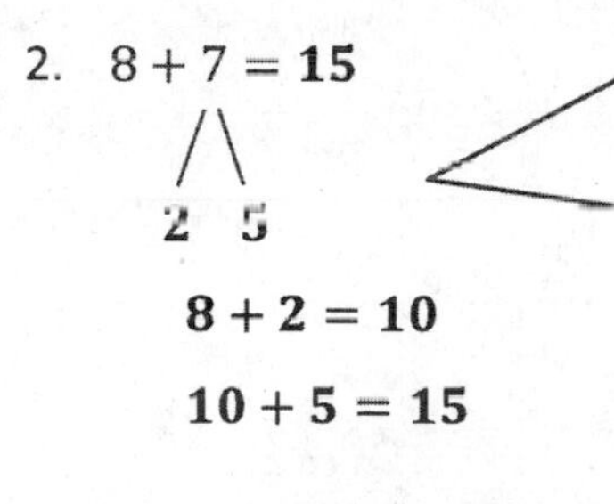

I can also solve without a drawing.
8 is closer to 10 than 7, so I can make 10 with the 8.
8 needs 2 to make 10, so I can break apart 7 with a number bond to get the 2 out.
Now I can add 8 and 2 to get 10, and now it is easy to add what is left; 10 and 5 is 15.
So 8 + 7 is 15.

3. 10 + **2** = 12

To solve, I can think 10 and what make 12? 10 and 2 make 12.

4. 9 + **3** = 12

I know 9 is 1 less than 10, so the answer for 9 + __ = 12 must be 1 more than 10 + __ = 12.
So 9 + 3 = 12.

5. Ronnie uses 5 brown bricks and 8 red bricks to build a fort. How many bricks does Ronnie use in all?

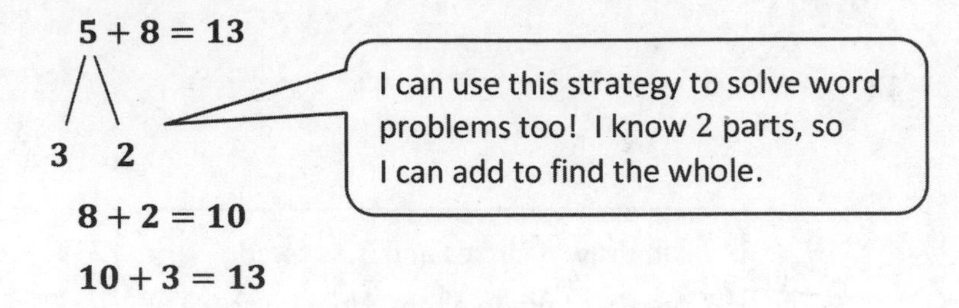

$8 + 2 = 10$

$10 + 3 = 13$

Ronnie used 13 bricks in all.

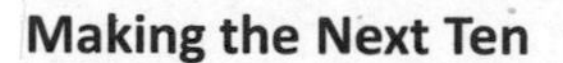

Making the Next Ten

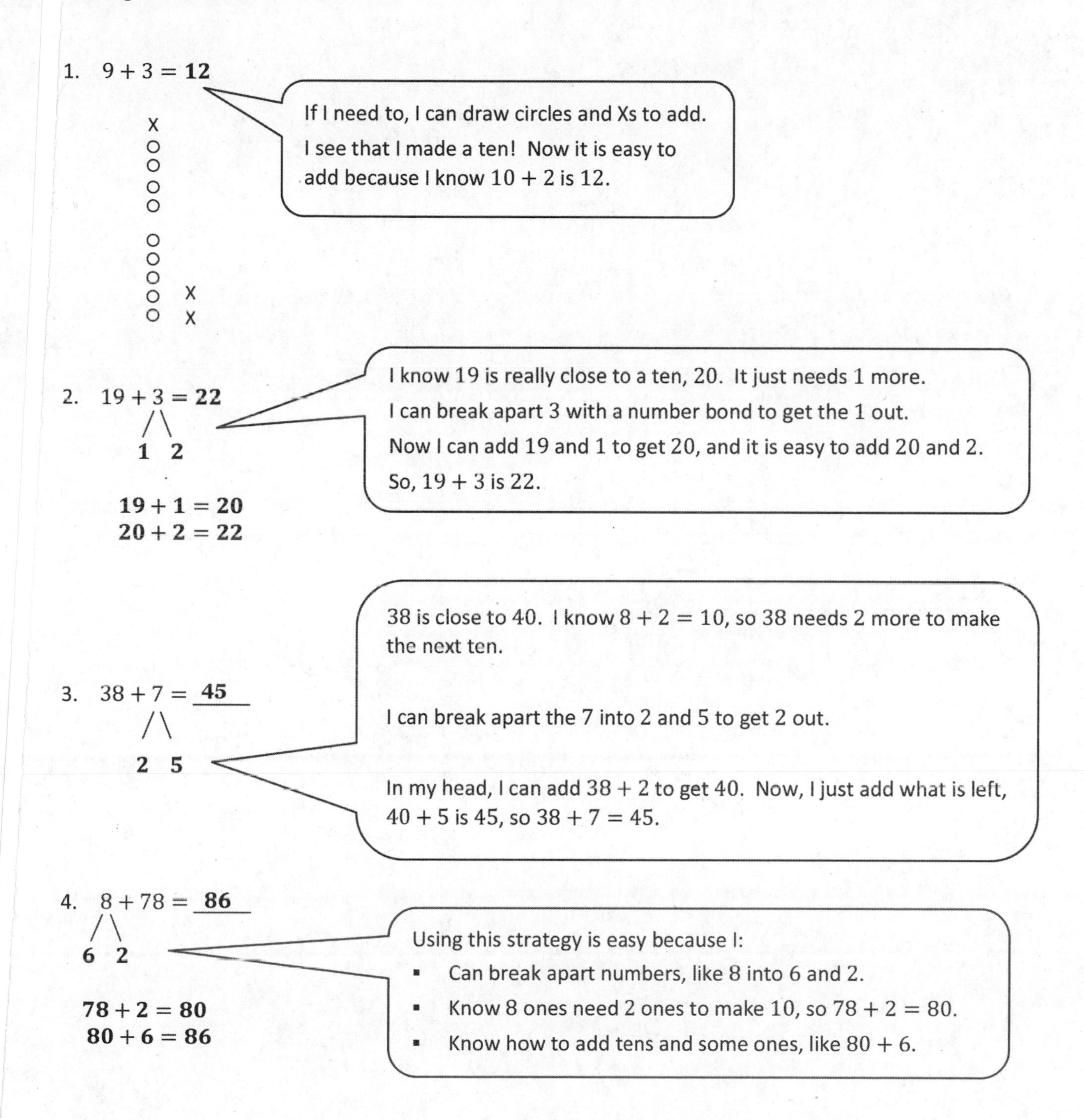

1. $9 + 3 = \mathbf{12}$

X
O
O
O
O
O
O
O
O X
O X

If I need to, I can draw circles and Xs to add.
I see that I made a ten! Now it is easy to add because I know $10 + 2$ is 12.

2. $19 + 3 = \mathbf{22}$
   (3 broken apart into **1** and **2**)

   $\mathbf{19 + 1 = 20}$
   $\mathbf{20 + 2 = 22}$

I know 19 is really close to a ten, 20. It just needs 1 more.
I can break apart 3 with a number bond to get the 1 out.
Now I can add 19 and 1 to get 20, and it is easy to add 20 and 2.
So, $19 + 3$ is 22.

3. $38 + 7 =$ __**45**__
   (7 broken apart into **2** and **5**)

38 is close to 40. I know $8 + 2 = 10$, so 38 needs 2 more to make the next ten.

I can break apart the 7 into 2 and 5 to get 2 out.

In my head, I can add $38 + 2$ to get 40. Now, I just add what is left, $40 + 5$ is 45, so $38 + 7 = 45$.

4. $8 + 78 =$ __**86**__
   (8 broken apart into **6** and **2**)

   $\mathbf{78 + 2 = 80}$
   $\mathbf{80 + 6 = 86}$

Using this strategy is easy because I:
- Can break apart numbers, like 8 into 6 and 2.
- Know 8 ones need 2 ones to make 10, so $78 + 2 = 80$.
- Know how to add tens and some ones, like $80 + 6$.

Name ______________________________ Date ______________

1. Solve.

a. 9 + 3 = ____ 1 2	b. 29 + 5 = ____
c. 49 + 7 = ____	d. 59 + 6 = ____
e. 18 + 4 = ____	f. 48 + 6 = ____
g. 58 + 6 = ____	h. 78 + 8 = ____

2. Solve.

a. 67 + 5 = ______	b. 87 + 6 = ______
c. 6 + 59 = ______	d. 7 + 78 = ______

3. Use the RDW process to solve.

There were 28 students at recess. A group of 7 students came outside to join them. How many students are there now?

____

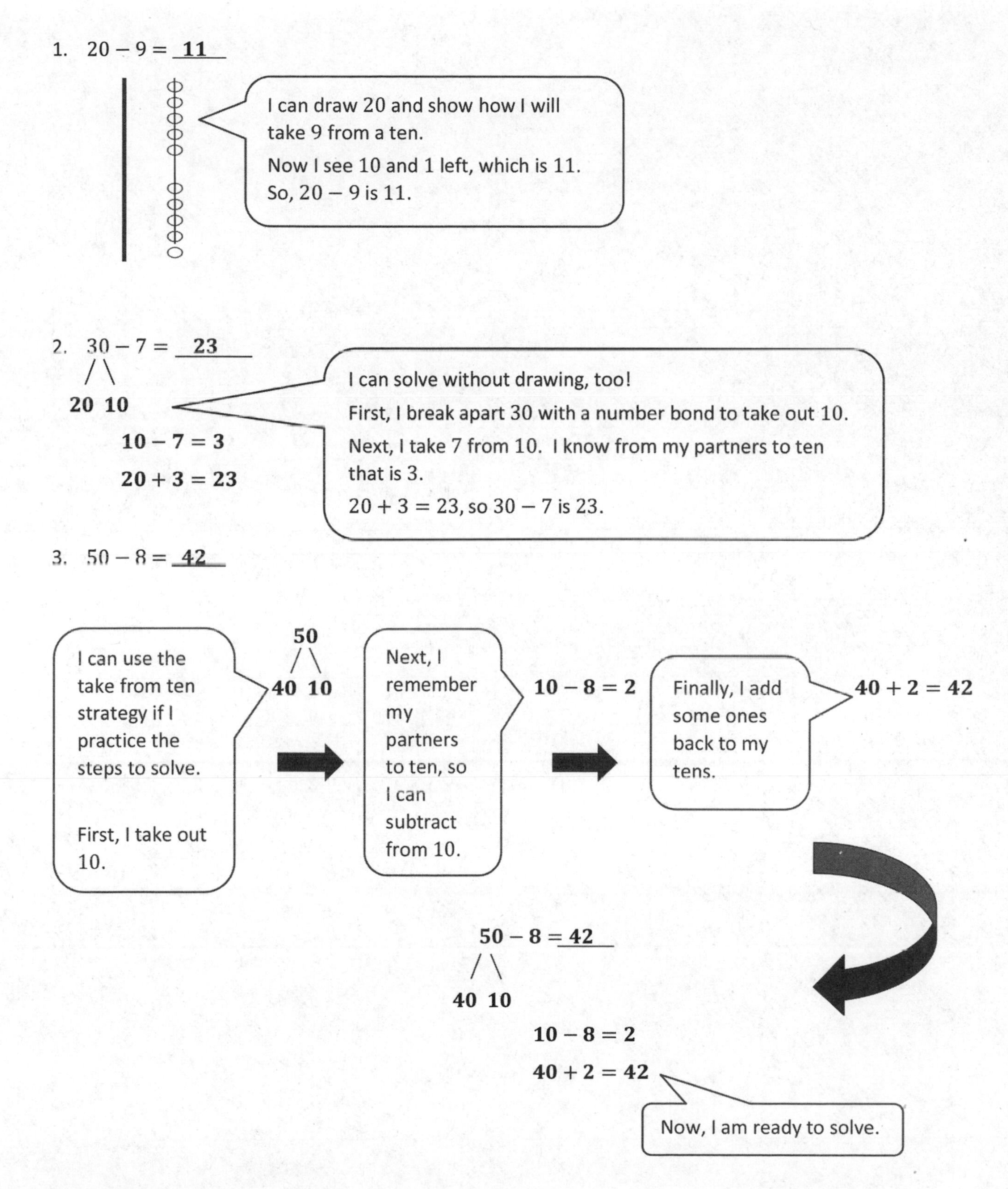
1. 20 − 9 = 11
I can draw 20 and show how I will take 9 from a ten.
Now I see 10 and 1 left, which is 11.
So, 20 − 9 is 11.
2. 30 − 7 = 23
20 10
10 − 7 = 3
20 + 3 = 23
I can solve without drawing, too!
First, I break apart 30 with a number bond to take out 10.
Next, I take 7 from 10. I know from my partners to ten that is 3.
20 + 3 = 23, so 30 − 7 is 23.
3. 50 − 8 = 42
I can use the take from ten strategy if I practice the steps to solve.
First, I take out 10.
50
40 10
Next, I remember my partners to ten, so I can subtract from 10.
10 − 8 = 2
Finally, I add some ones back to my tens.
40 + 2 = 42
50 − 8 = 42
40 10
10 − 8 = 2
40 + 2 = 42
Now, I am ready to solve.

Take from 10

1. $12 - 9 = \mathbf{3}$

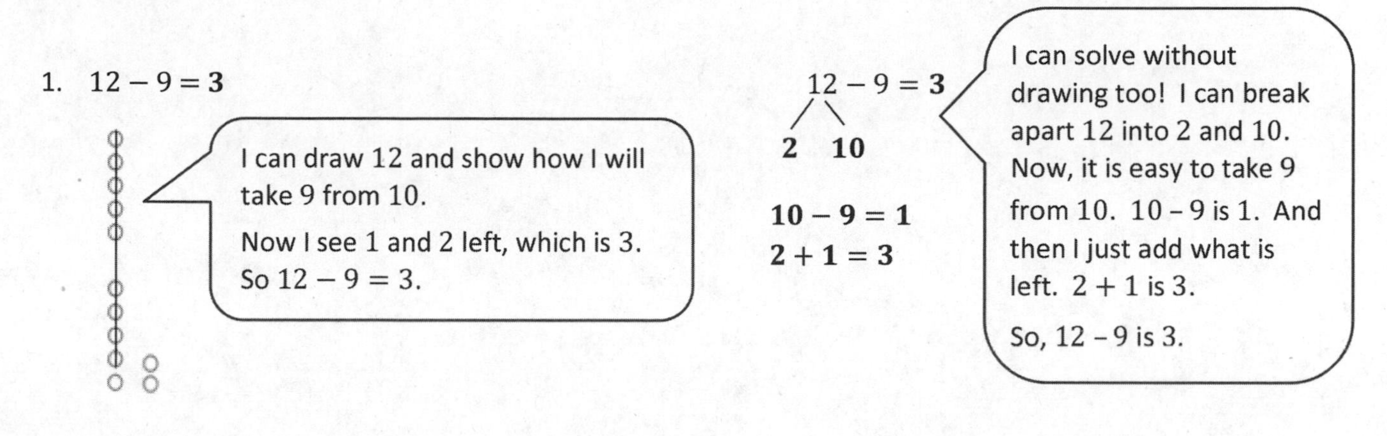

2. $14 - 8 = \mathbf{6}$

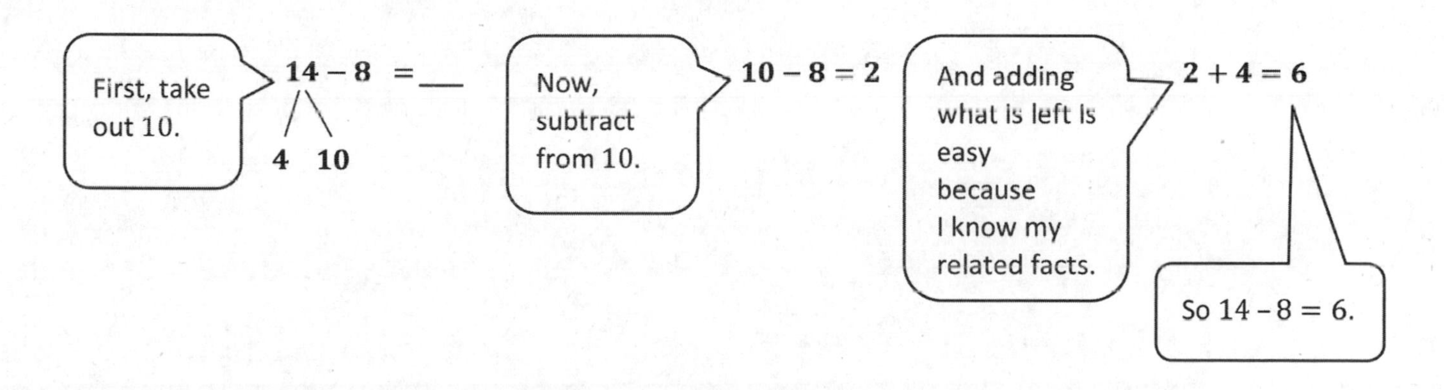

3. Shane has 12 pencils. He gives some pencils to his friends. Now, he has 7 left. How many pencils did he give away?

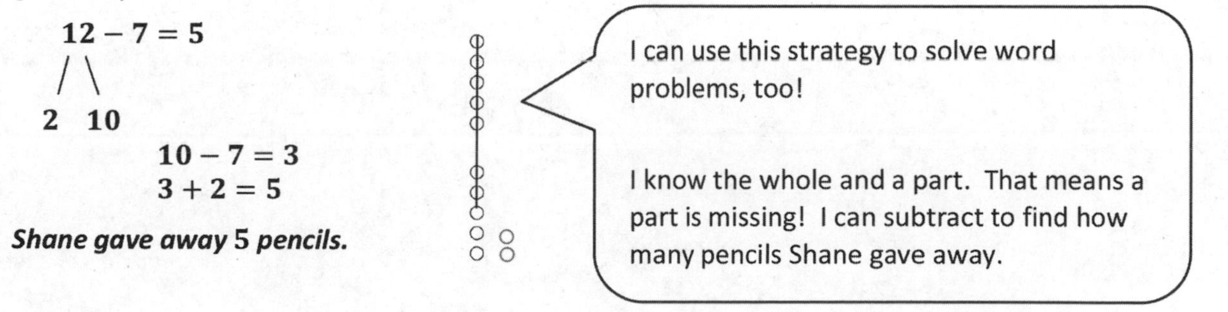

Name ________________________________ Date ________________

1. Take out ten.

17 / \ 7 10	14	18
13	16	19

2. Solve.

10 - 2 = _____	10 - 7 = _____	10 - 6 = _____
10 - 5 = _____	10 - 8 = _____	10 - 9 = _____

3. Solve.

a. 14 - 9 = _____ /\ 4 10 10 - 9 = 1 1 + 4 = _____	b. 15 - 8 = _____
c. 13 - 7 = _____	d. 12 - 8 = _____

Solve.

4. Robert has 16 cups. Some are red. Nine are blue. How many cups are red?

_____ cups are red.

5. Lucy spent $8 on a game. She started with $14. How much money does Lucy have left?

____

Take from 10

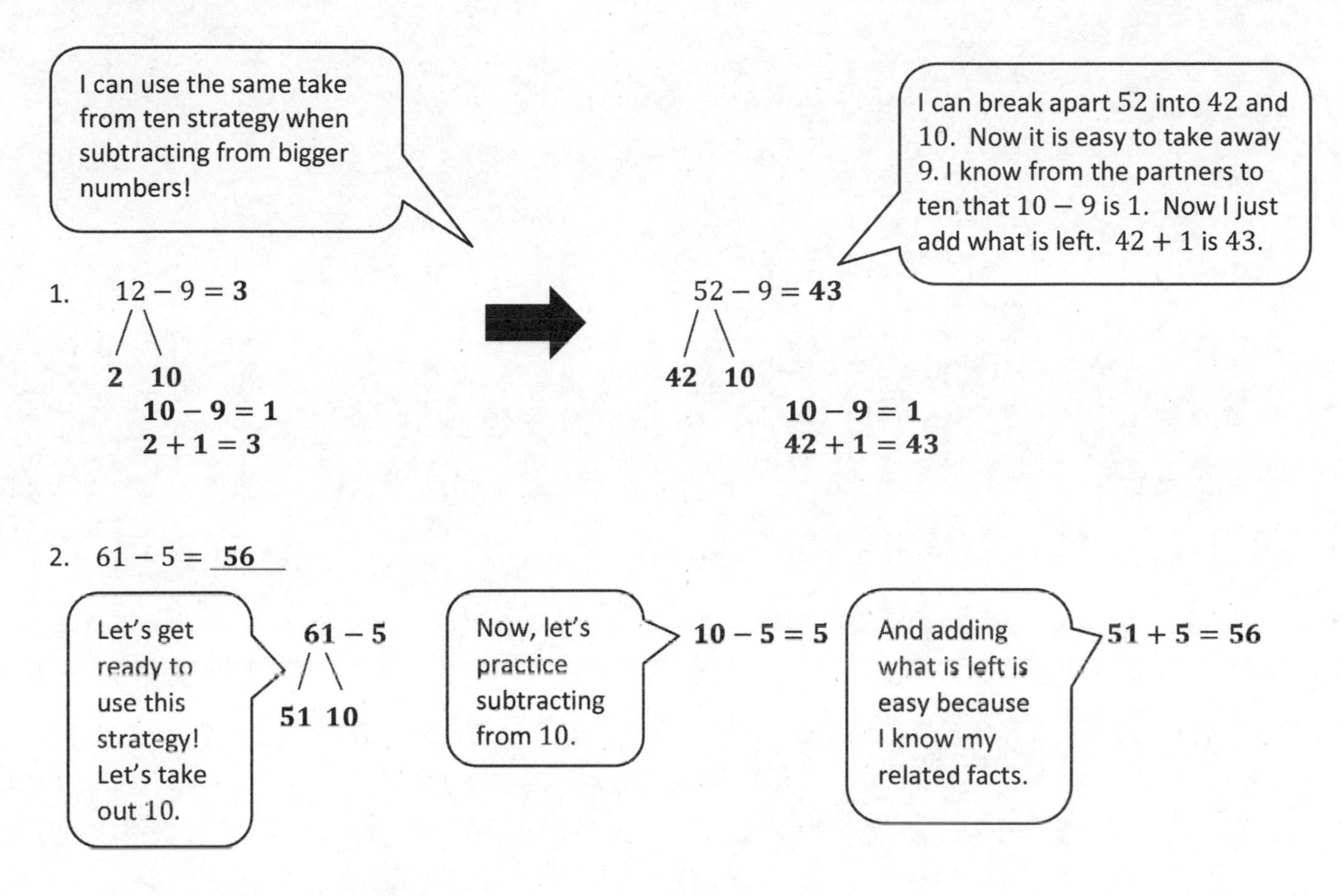

I can use the same take from ten strategy when subtracting from bigger numbers!

I can break apart 52 into 42 and 10. Now it is easy to take away 9. I know from the partners to ten that $10 - 9$ is 1. Now I just add what is left. $42 + 1$ is 43.

1. $12 - 9 = \mathbf{3}$

   2 10

   $\mathbf{10 - 9 = 1}$

   $\mathbf{2 + 1 = 3}$

   $52 - 9 = \mathbf{43}$

   42 10

   $\mathbf{10 - 9 = 1}$

   $\mathbf{42 + 1 = 43}$

2. $61 - 5 =$ **56**

   Let's get ready to use this strategy! Let's take out 10.

   61 − 5

   51 10

   Now, let's practice subtracting from 10.

   $\mathbf{10 - 5 = 5}$

   And adding what is left is easy because I know my related facts.

   $\mathbf{51 + 5 = 56}$

3. Mrs. Watts had 12 tacos. The children ate some. Nine tacos were left. How many tacos did the children eat?

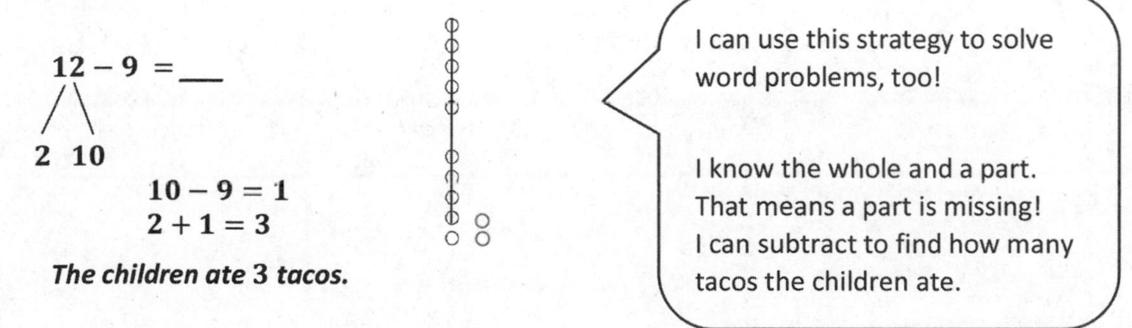

$\mathbf{12 - 9 =}$ ___

2 10

$\mathbf{10 - 9 = 1}$

$\mathbf{2 + 1 = 3}$

The children ate 3 tacos.

I can use this strategy to solve word problems, too!

I know the whole and a part. That means a part is missing! I can subtract to find how many tacos the children ate.

Name ______________________________ Date ______________

1. Take out ten.

26 / \\ 16 10	34	58
85	77	96

2. Solve.

10 – 1 = _____	10 – 5 = _____	10 – 2 = _____
10 – 4 = _____	10 – 7 = _____	10 – 8 = _____

3. Solve.

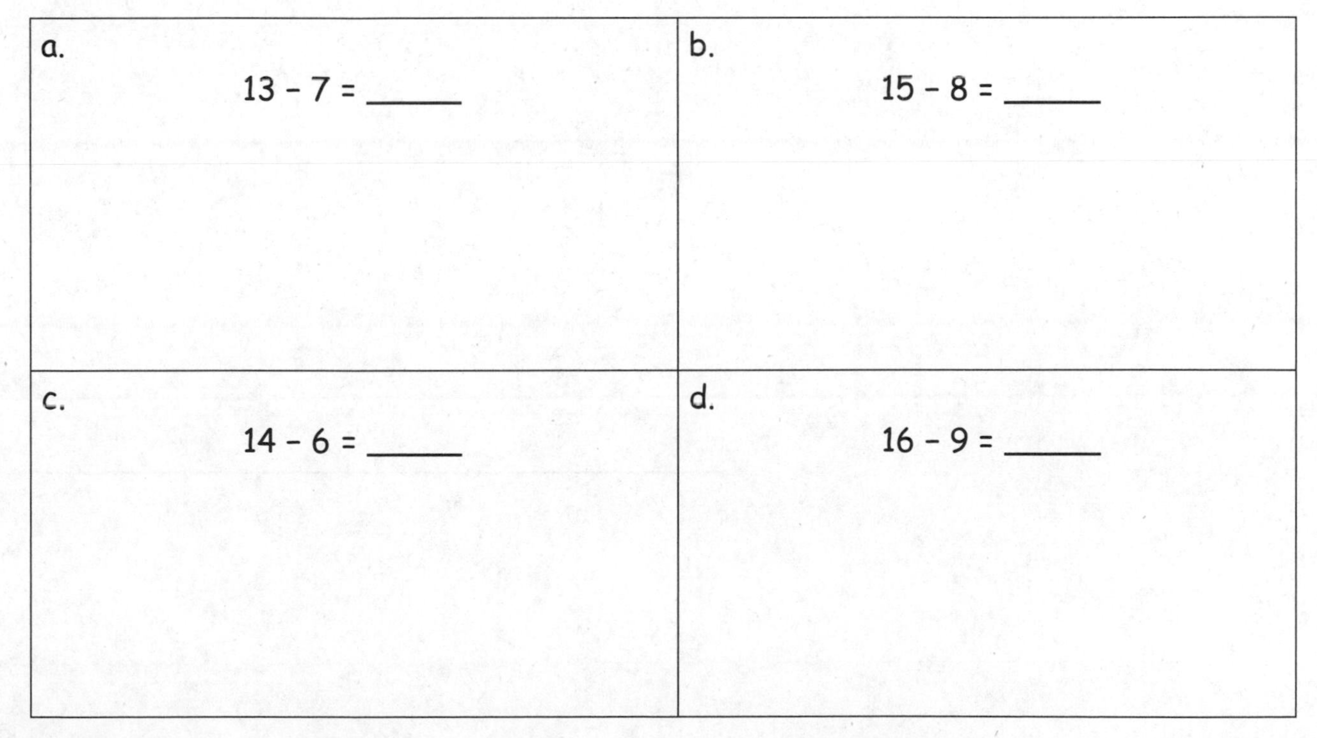

a. 13 – 7 = _____	b. 15 – 8 = _____
c. 14 – 6 = _____	d. 16 – 9 = _____

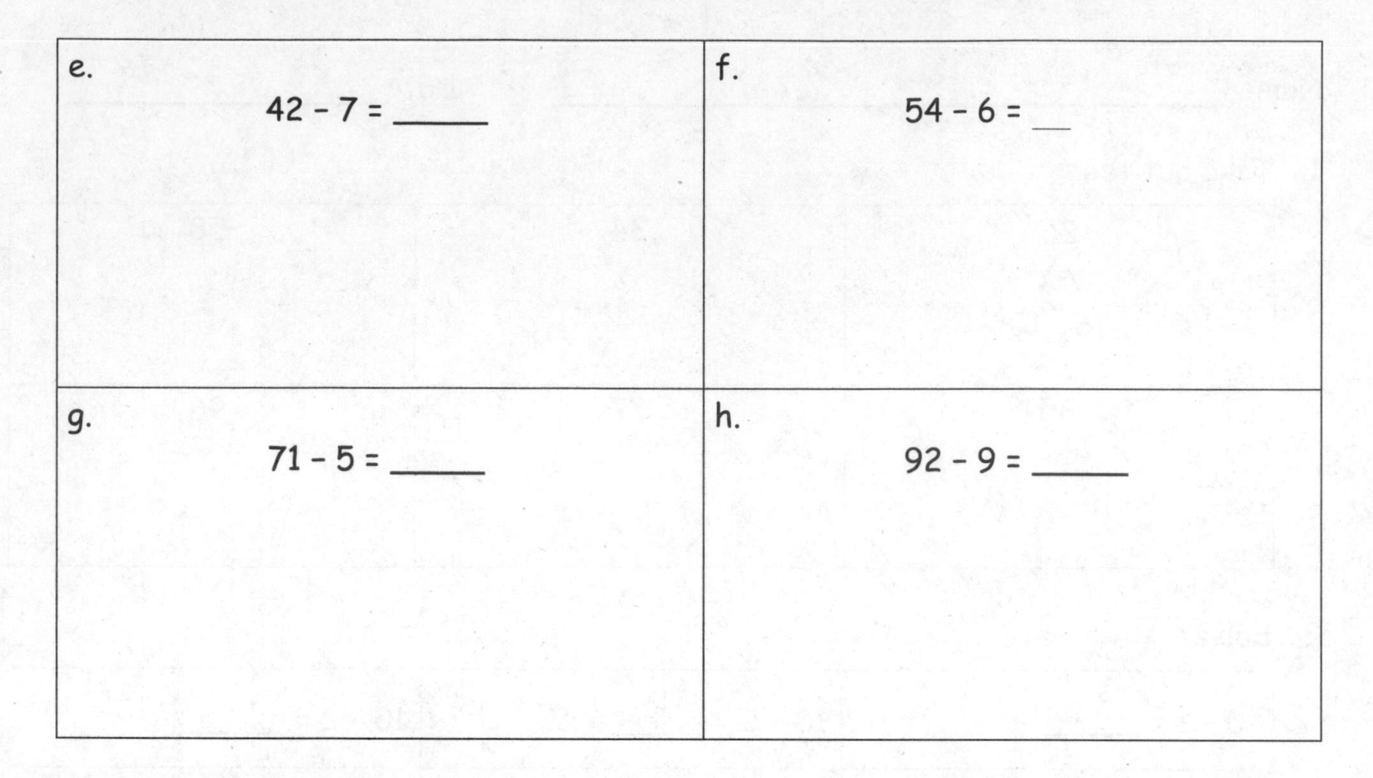

e. 42 – 7 = ______	f. 54 – 6 = ___
g. 71 – 5 = ______	h. 92 – 9 = ______

4. Emma has 16 markers. She gave Jack some. Seven markers are left. How many markers did Emma give Jack?

____

Grade 2
Module 2

1. The length of the picture of the shovel is about __8__ centimeters.

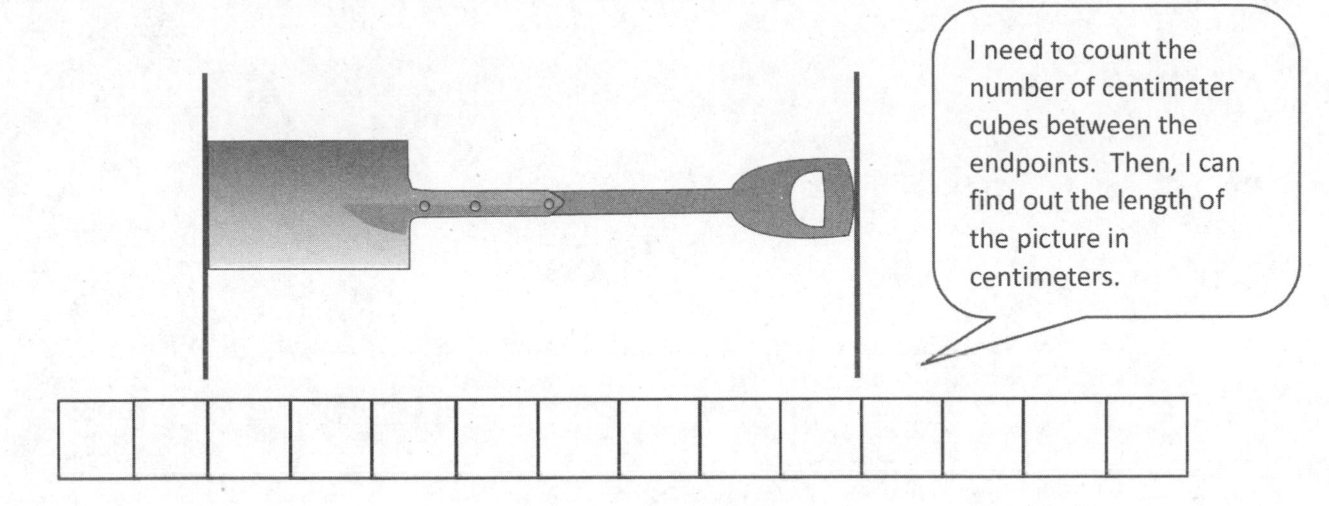

2. The length of a screwdriver is 19 centimeters. The handle is 5 centimeters long. What is the length of the top of the screwdriver?

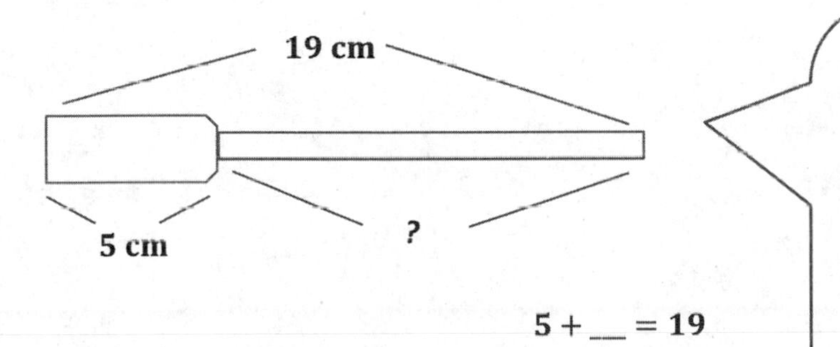

I can use the Read-Draw-Write process to solve. I can draw a screwdriver and label the whole length 19 cm. This is just like lining up my centimeter cubes! I know one part is 5 centimeters, so I'll label that. I can use addition to solve for the unknown part, which is 14 cm. I can write a complete statement of my answer.

The top of the screwdriver is 14 centimeters.

Name ______________________________ Date ____________

Count each centimeter cube to find the length of each object.

1. The crayon is ________ centimeter cubes long.

2. The pencil is ________ centimeter cubes long.

3. The clothespin is ________ centimeter cubes long.

4. The length of the marker is ______ centimeter cubes.

5. Richard has 43 centimeter cubes. Henry has 30 centimeter cubes. What is the length of their cubes altogether?

6. The length of Marisa's loaf of bread is 54 centimeters. She cut off and ate 7 centimeters of bread. What is the length of what she has left?

7. The length of Jimmy's math book is 17 centimeter cubes. His reading book is 12 centimeter cubes longer. What is the length of his reading book?

1. The picture of the eraser is about **4** centimeters long.

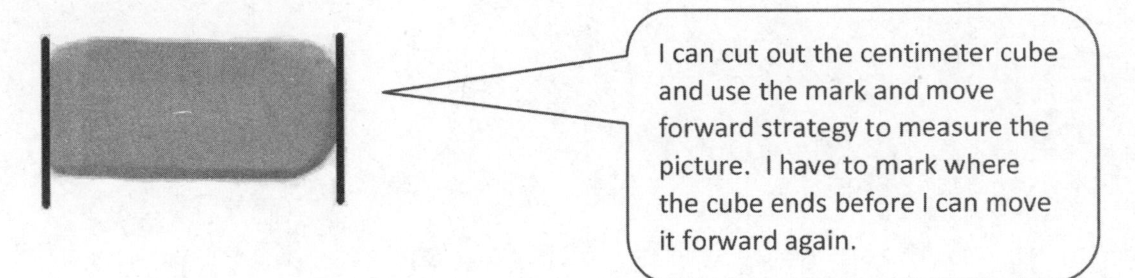

2. John used a centimeter cube and the mark and move forward strategy to measure these pieces of tape. Use his work to answer the following questions.

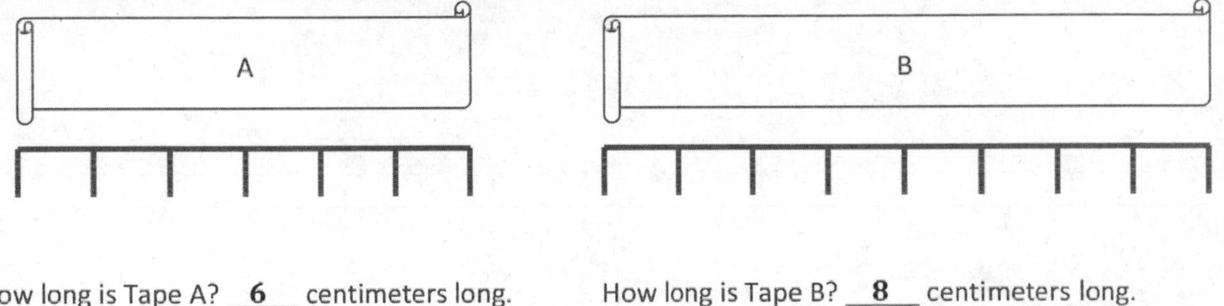

How long is Tape A? **6** centimeters long.

How long is Tape B? **8** centimeters long.

Which tape is shorter? ***Tape A***

The total length of Tapes A and B is **14** centimeters.

> Since John measured without any gaps or overlaps, I know that the distance between the pencil marks is the same length! I can count the length units for each piece of tape.

Name ______________________________ Date ______________

Use the centimeter square at the bottom of the next page to measure the length of each object. Mark the endpoint of the square as you measure.

1. The picture of the glue is about ________ centimeters long.

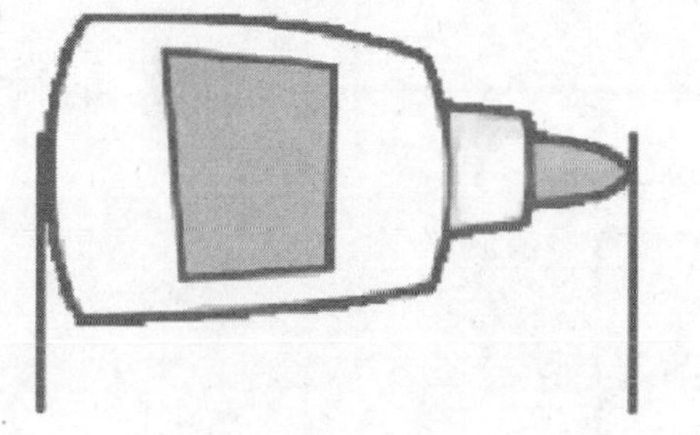

2. The picture of the lollipop is about _______ centimeters long.

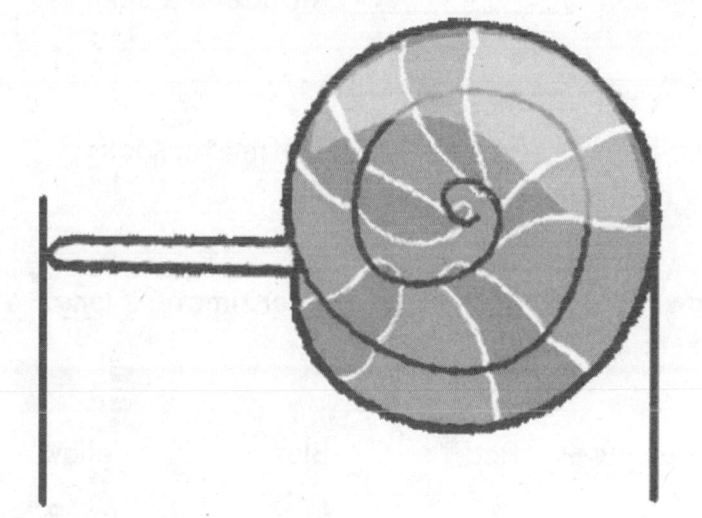

3. The picture of the scissors is about _______ centimeters long.

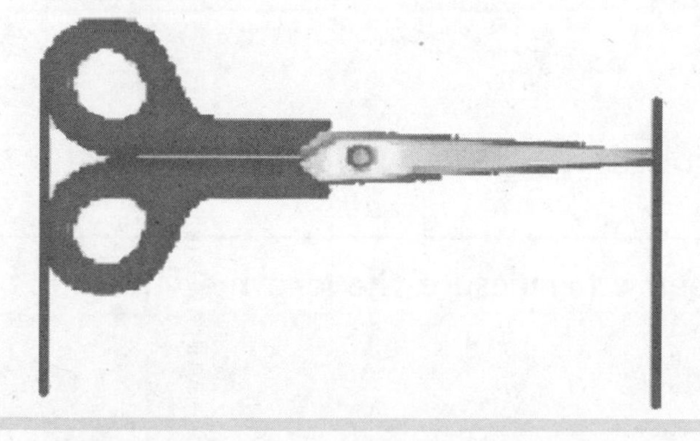

4. Samantha used a centimeter cube and the mark and move forward strategy to measure these ribbons. Use her work to answer the following questions.

Red Ribbon

Blue Ribbon

Yellow Ribbon

a. How long is the red ribbon? ___________ centimeters long.

b. How long is the blue ribbon? ___________ centimeters long.

c. How long is the yellow ribbon? ___________ centimeters long.

d. Which ribbon is the longest? Red Blue Yellow

e. Which ribbon is the shortest? Red Blue Yellow

f. The total length of the ribbons is ___________ centimeters.

Cut out the centimeter square below to measure the length of the glue bottle, lollipop, and scissors.

Use your centimeter ruler to answer the following questions.

1. The picture of the animal track is about __4__ cm long.

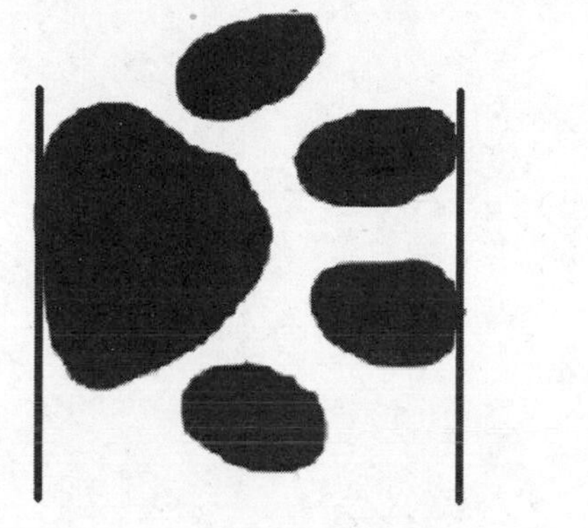

I know how to accurately line up my centimeter ruler to measure the picture of the animal track. Since my hash marks are labeled, I don't have to count each mark; I can easily see that the picture is 4 centimeters long.

2. Measure the lengths of sides A, B, and C. Write each length on the line.

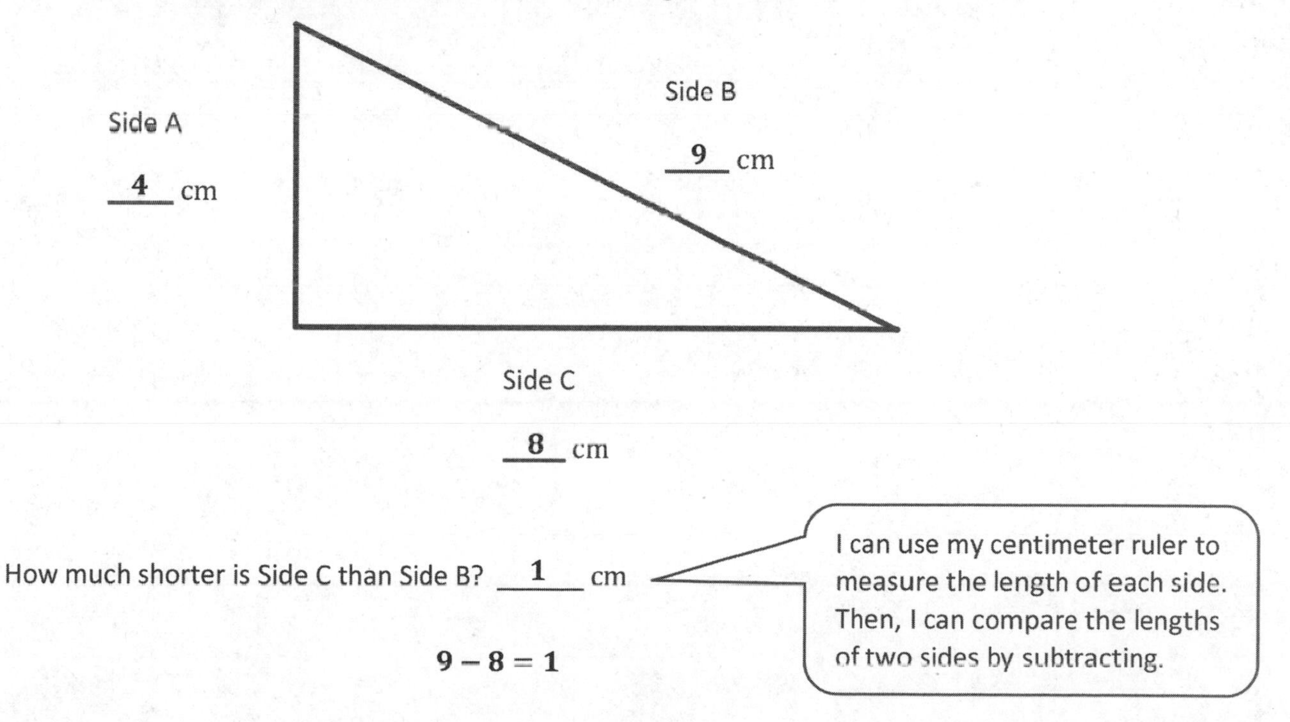

How much shorter is Side C than Side B? __1__ cm

$9 - 8 = 1$

I can use my centimeter ruler to measure the length of each side. Then, I can compare the lengths of two sides by subtracting.

Name ______________________________ Date ______________

Measure the lengths of the objects with the centimeter ruler you made in class.

1. The picture of the fish is ____ cm long.

2. The picture of the fish tank is _________ cm long.

© ciroorabona – Fotolia.com

3. The picture of the fish tank is ________ cm longer than the picture of the fish.

4. Measure the lengths of Sides A, B, and C. Write each length on the line.

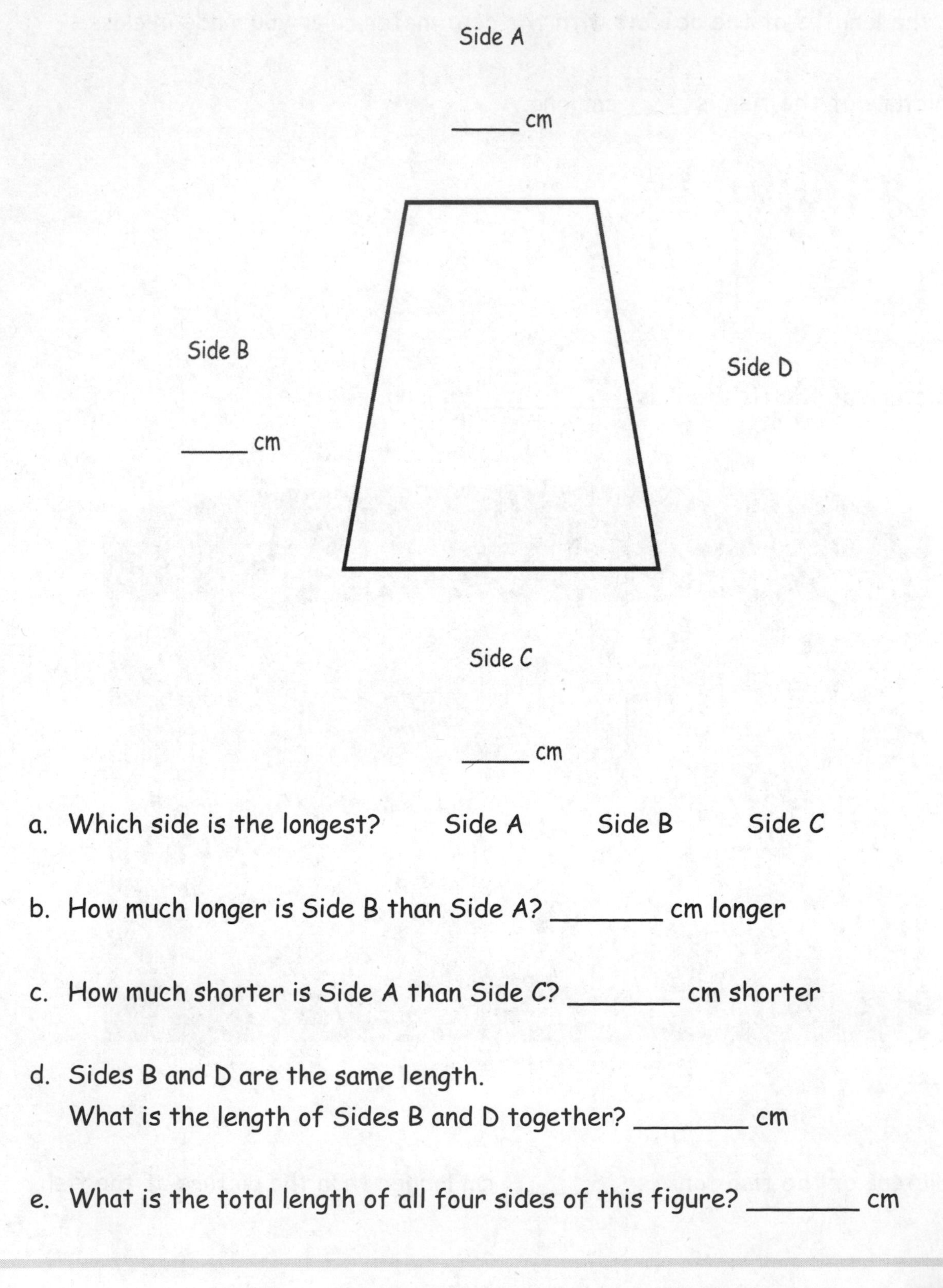

a. Which side is the longest? Side A Side B Side C

b. How much longer is Side B than Side A? _______ cm longer

c. How much shorter is Side A than Side C? _______ cm shorter

d. Sides B and D are the same length.
What is the length of Sides B and D together? _______ cm

e. What is the total length of all four sides of this figure? _______ cm

1. Circle cm (centimeter) or m (meter) to show which unit you would use to measure the length of each object.

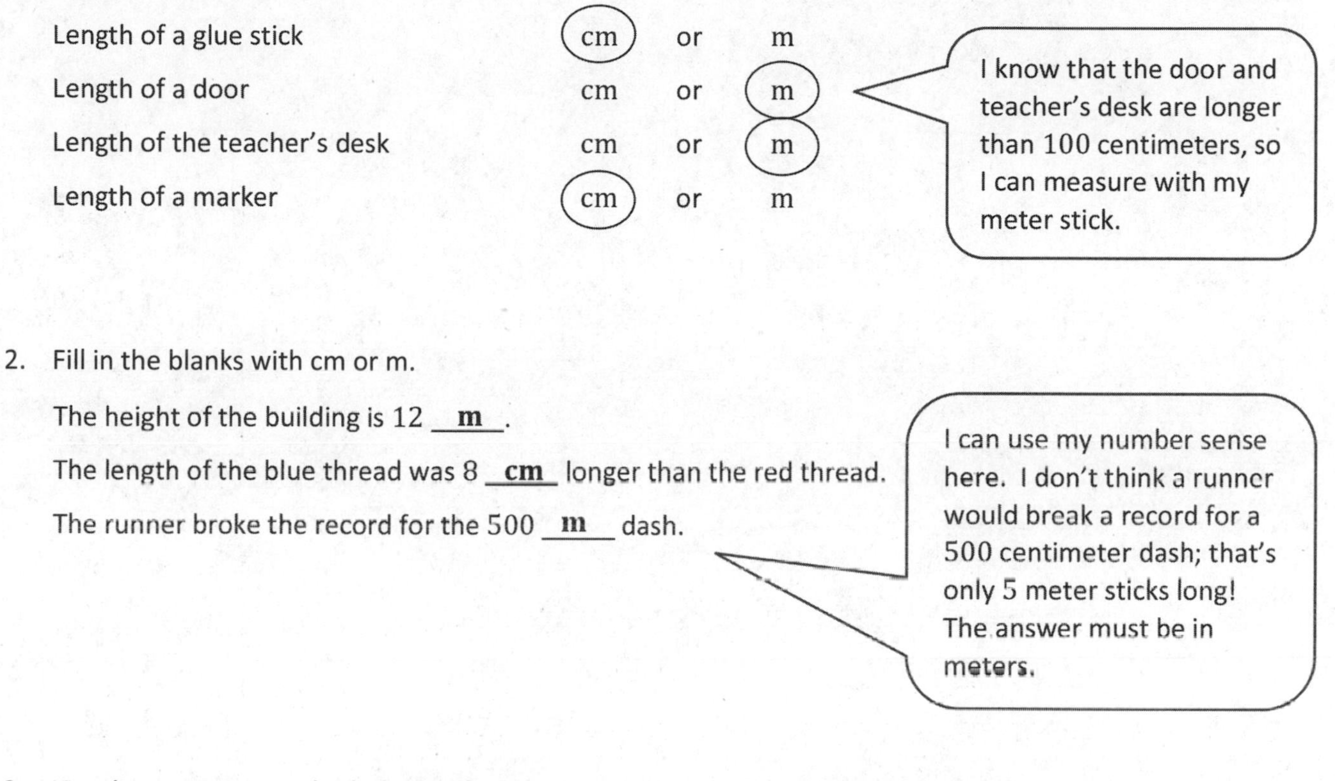

3. Use the centimeter ruler below to find the length (from one mark to the next) of the shape.

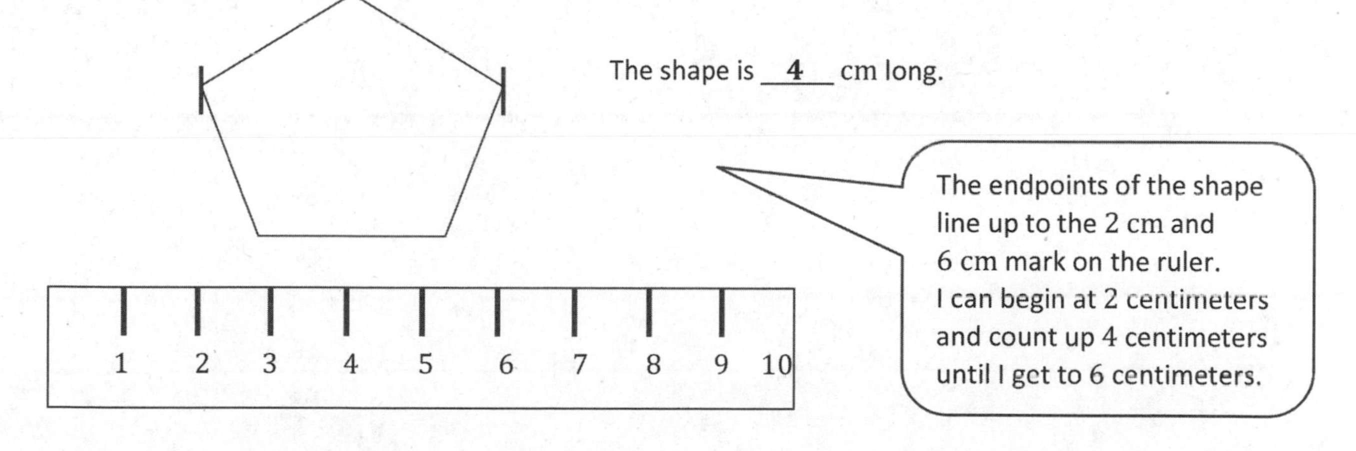

Name ______________________________ Date ______________

1. Circle cm (centimeter) or m (meter) to show which unit you would use to measure the length of each object.

   a. Length of a marker cm or m

   b. Length of a school bus cm or m

   c. Length of a laptop computer cm or m

   d. Length of a highlighter marker cm or m

   e. Length of a football field cm or m

   f. Length of a parking lot cm or m

   g. Length of a cell phone cm or m

   h. Length of a lamp cm or m

   i. Length of a supermarket cm or m

   j. Length of a playground cm or m

2. Fill in the blanks with **cm** or **m**.

   a. The length of a swimming pool is 25 ____________.

   b. The height of a house is 8 ____________.

   c. Karen is 6 ____________ shorter than her sister.

   d. Eric ran 65 ____________ down the street.

   e. The length of a pencil box is 3 ____________ longer than a pencil.

3. Use the centimeter ruler to find the length (from one mark to the next) of each object.

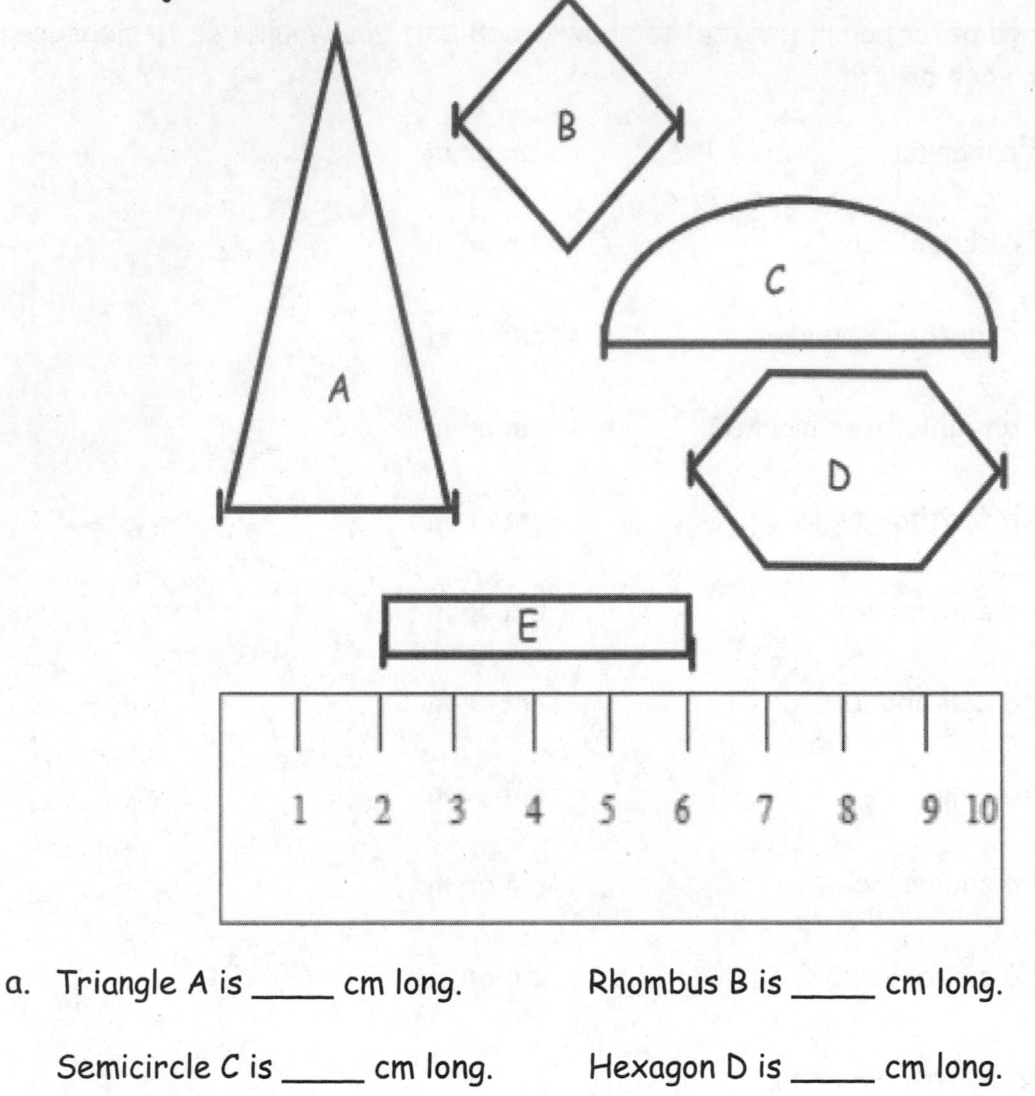

a. Triangle A is ____ cm long. Rhombus B is ____ cm long.

Semicircle C is ____ cm long. Hexagon D is ____ cm long.

Rectangle E is ____ cm long.

b. Explain how the strategy to find the length of each shape above is different from how you would find the length if you used a centimeter cube.

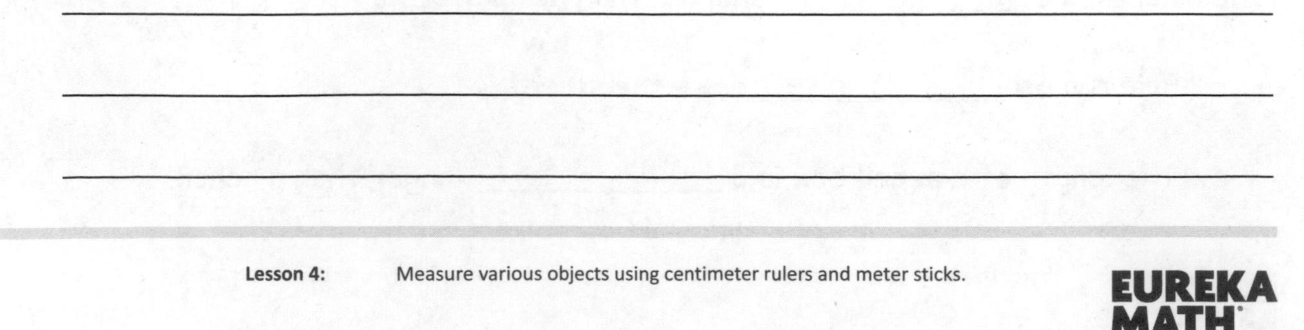

1. Name two things in school that you would measure in meters. Estimate their lengths.

Item	Estimated Length
chalkboard	***4 meters***
reading corner rug	***3 meters***

I know that the length from the doorknob to the floor is about 1 meter. So I think the reading corner rug is about 3 of those lengths. The rug looks shorter than the chalkboard, so I can estimate that the rug is about 3 meters long.

2. Choose the best length estimate for each object.

   a. Bulletin board (2 m) or 35 cm

   b. Scissors (13 cm) or 43 cm

   c. Top of a student desk 18 cm or (62 cm)

I know that a 3-ring binder is about 30 centimeters long. I can picture 2 of those binders fitting across the length of my desktop, which would be about 60 centimeters long. So, 62 centimeters is closer to 60 centimeters than 18 centimeters.

3. Measure the length of the line below using your pinky finger. Write your estimate.

Estimate: ___7___ cm

Since the width of my pinky finger is about 1 centimeter, I can estimate that the length of the line is about 7 centimeters.

Name _________________________ Date _______________

1. Name five things in your home that you would measure in meters.
   Estimate their length.
   *Remember, the length from a doorknob to the floor is about 1 meter.

Item	Estimated Length
a.	
b.	
c.	
d.	
e.	

2. Choose the best length estimate for each object.

a. Whiteboard	3 m	or	45 cm
b. Banana	14 cm	or	30 cm
c. DVD	25 cm	or	17 cm
d. Pen	16 cm	or	1 m
e. Swimming pool	50 m	or	150 cm

3. The width of your pinky finger is about 1 cm.
   Measure the length of the lines using your pinky finger. Write your estimate.

   a. Line A ______________________________

      Line A is about __________ cm long.

   b. Line B _____

      Line B is about __________ cm long.

   c. Line C

      ____

      Line C is about __________ cm long.

   d. Line D ____

      Line D is about __________ cm long.

   e. Line E ____________________

      Line E is about __________ cm long.

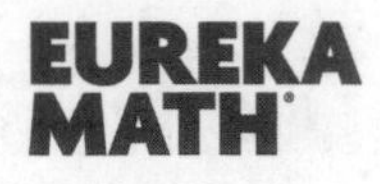

1. Measure each set of lines in centimeters, and write the length on the line. Complete the comparison sentences.

Line A

Line B

Line C

> I can lay my meter strip along each line to measure its length. I need to line up the zero point on my ruler with the endpoint of the line!

Line A Line B Line C

15 cm **5** cm **8** cm

Lines A, B, and C are about **28** cm combined.

Line C is about **7** cm shorter than Line A.

> Since Line A is 15 cm long and Line C is 8 cm long, I know that Line C is shorter. I can subtract: $15 - 8 = 7$. Line C is 7 cm shorter than Line A.

2. Line D is 45 cm long. Line E is 70 cm long. Line F is 1 m long.

Line E is **25** cm longer than Line D.

Line E doubled is **40** cm longer than Line F.

> I know that 1 meter equals 100 centimeters. If I double Line E, then it will be 140 cm long because $70 + 70 = 140$. 140 centimeters is 40 centimeters more than 100 centimeters.

3. Lanie measured the height of her little brother. He is 52 cm tall.

How much taller is a meter stick than her brother? **48** cm.

$$52 + \_\_ = 100$$
$$52 + 8 = 60$$
$$60 + 40 = 100$$
$$8 + 40 = 48$$

> This is like a missing addend problem. I can solve by adding on. I want to get to 100 because a meter stick is 100 cm long. I know that $52 + 8$ will get me to the friendly number 60. Then, $60 + 40 = 100$. And, $8 + 40 = 48$.

Name ____________________________ Date ______________

Measure each set of lines in centimeters, and write the length on the line. Complete the comparison sentences.

1. Line A

   Line B

   a. Line A is about ________ cm longer than line B.

   b. Line A and B are about ________ cm combined.

2. Line X

   Line Y

   Line Z

   a. Line X ______ cm   Line Y ______ cm   Line Z ______ cm

   b. Lines X, Y, and Z are about ________ cm combined.

   c. Line Z is about ________ cm shorter than Line X.

   d. Line X is about ________ cm shorter than Line Y.

   e. Line Y is about ________ cm longer than Line Z.

   f. Line X doubled is about ________ cm longer than line Y.

3. Line J is 60 cm long. Line K is 85 cm long. Line L is 1 m long.

   a. Line J is ________ cm shorter than line K.

   b. Line L is ________ cm longer than line K.

   c. Line J doubled is ________ cm more than line L.

   d. Lines J, K, and L combined are ________ cm.

4. Katie measured the seat height of four different chairs in her house. Here are her results:

   Loveseat height: 51 cm
   Dining room chair height: 55 cm
   Highchair height: 97 cm
   Counter stool height: 65 cm

   a. How much shorter is the dining room chair than the counter stool? ________ cm

   b. How much taller is a meter stick than the counter stool? ________ cm

   c. How much taller is a meter stick than the loveseat? ________ cm

5. Max ran 15 meters this morning. This afternoon, he ran 48 meters.

   a. How many more meters did he run in the afternoon?

   b. How many meters did Max run in all?

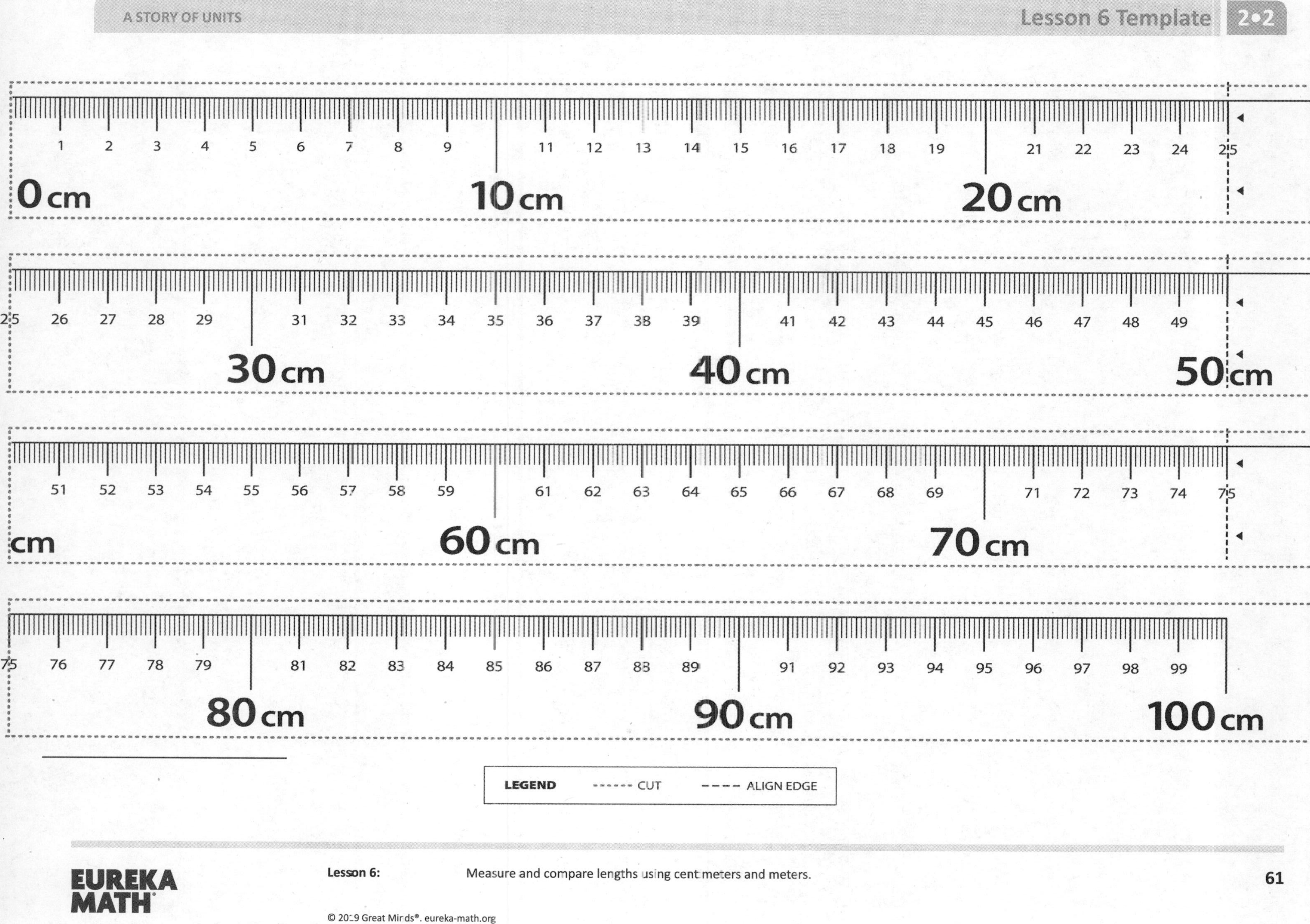
0 cm
10 cm
20 cm
30 cm
40 cm
50 cm
60 cm
70 cm
80 cm
90 cm
100 cm
LEGEND
------ CUT
---- ALIGN EDGE

1. Measure each line with one small paper clip, using the mark and move forward method. Then, measure in centimeters using a ruler.

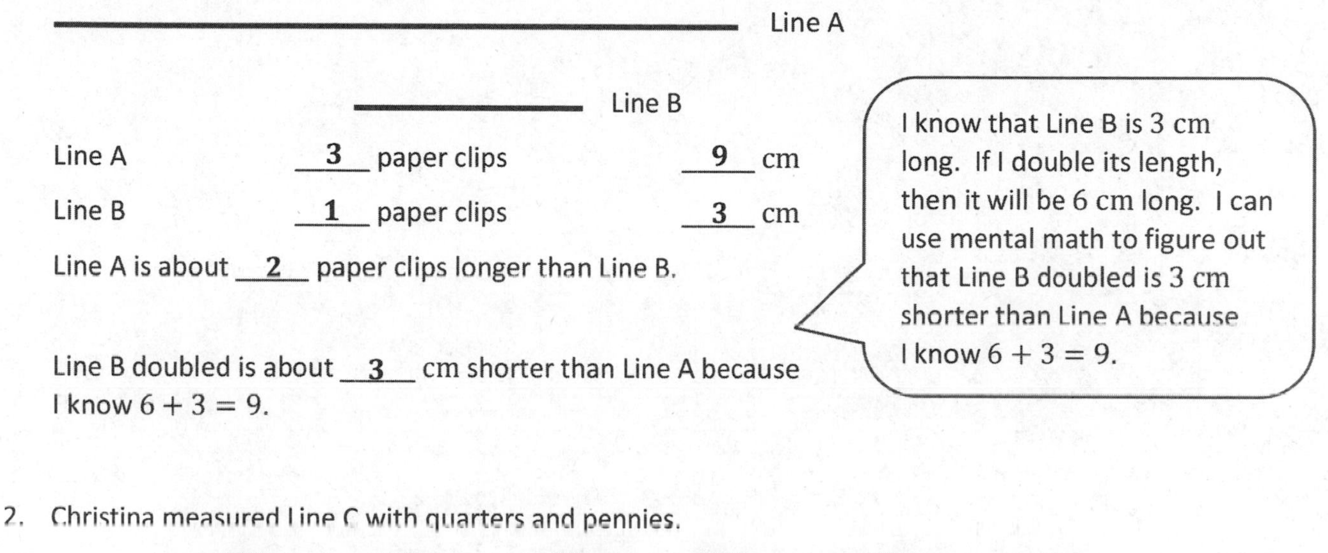

Line A

Line B

Line A ___3___ paper clips ___9___ cm

Line B ___1___ paper clips ___3___ cm

Line A is about ___2___ paper clips longer than Line B.

Line B doubled is about ___3___ cm shorter than Line A because I know $6 + 3 = 9$.

I know that Line B is 3 cm long. If I double its length, then it will be 6 cm long. I can use mental math to figure out that Line B doubled is 3 cm shorter than Line A because I know $6 + 3 = 9$.

2. Christina measured Line C with quarters and pennies.

Line C

Why did Christina need more pennies than quarters to measure Line C?

Since the quarter is bigger, it takes fewer quarters to measure the same line. If the length unit is smaller, like a penny, then you need a greater number of pennies to measure the line.

If the unit size is bigger, like quarters, then you need fewer units. If the unit size is smaller, like pennies, then you need more units. Coins aren't a good measurement tool. Centimeters are much more reliable because each length unit is the same!

Name ______________________________ Date ______________

Use a centimeter ruler and paper clips to measure and compare lengths.

1. ______________________________ Line Z

a. Line Z

_____ paper clips _____ cm

b. Line Z doubled would measure about _____ paper clips or about _____ cm long.

2. ____ Line A

____________________ Line B

a. Line A

_____ paper clips _____ cm

b. Line B

_____ paper clips _____ cm

c. Line A is about _____ paper clips longer than Line B.

b. Line B doubled is about _____ cm shorter than Line A.

3. Draw a line that is 9 cm long and another line below it that is 12 cm long.

Label the 9 cm line F and the 12 cm line G.

a. Line F   Line G

_____ paper clips   _____ paper clips

b. Line G is about _____ cm longer than Line F.

c. Line F is about _____ paper clips shorter than Line G.

d. Lines F and G are about _____ paper clips long.

e. Lines F and G are about _____ centimeters long

4. Jordan measured the length of a line with large paper clips. His friend measured the length of the same line with small paper clips.

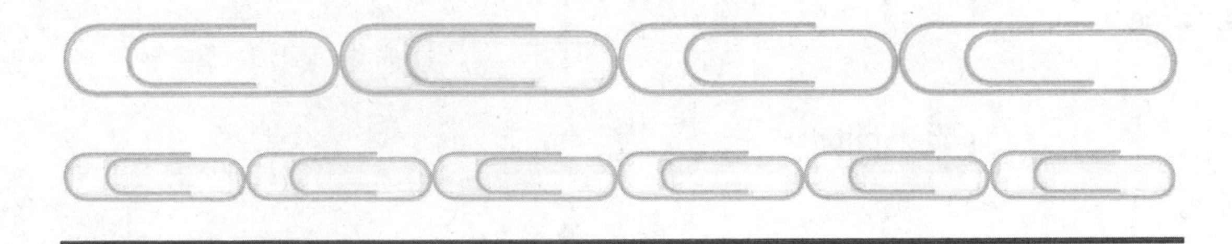

a. About how many paper clips did Jordan use? _____ large paper clips

b. About how many small paper clips did his friend use? _____ small paper clips

c. Why did Jordan's friend need more paper clips to measure the same line as Jordan?

____

____

1.

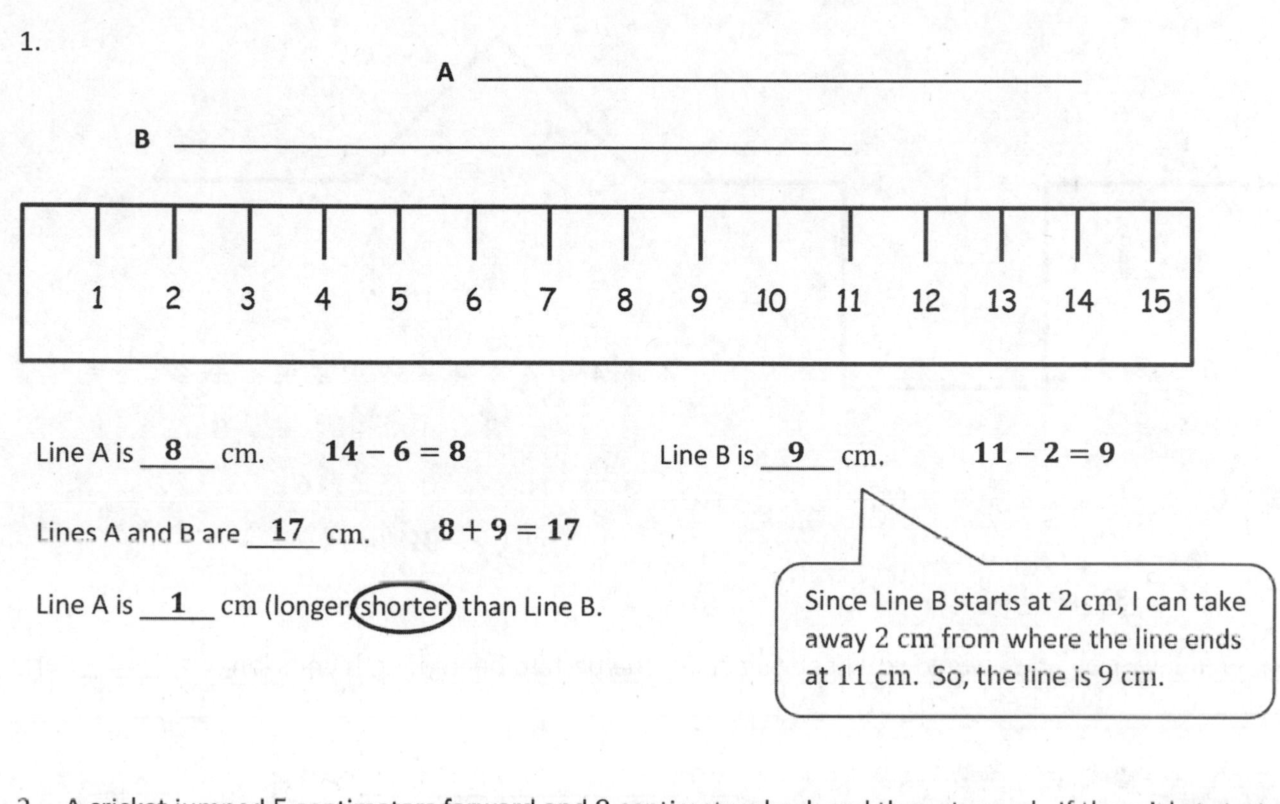

Line A is **8** cm. $14 - 6 = 8$

Line B is **9** cm. $11 - 2 = 9$

Lines A and B are **17** cm. $8 + 9 = 17$

Line A is **1** cm (longer, **shorter**) than Line B.

Since Line B starts at 2 cm, I can take away 2 cm from where the line ends at 11 cm. So, the line is 9 cm.

2. A cricket jumped 5 centimeters forward and 9 centimeters back and then stopped. If the cricket started at 23 on the ruler, where did the cricket stop? Show your work on the broken centimeter ruler.

$23 + 5 = 28$

$28 - 9 = 18 + 1 = 19$ (28 split into 18 and 10)

I can use addition and subtraction to solve. I can start at 23 and count on 5. Then, I can hop back 9 centimeters or subtract 9. The cricket stops at 19 cm.

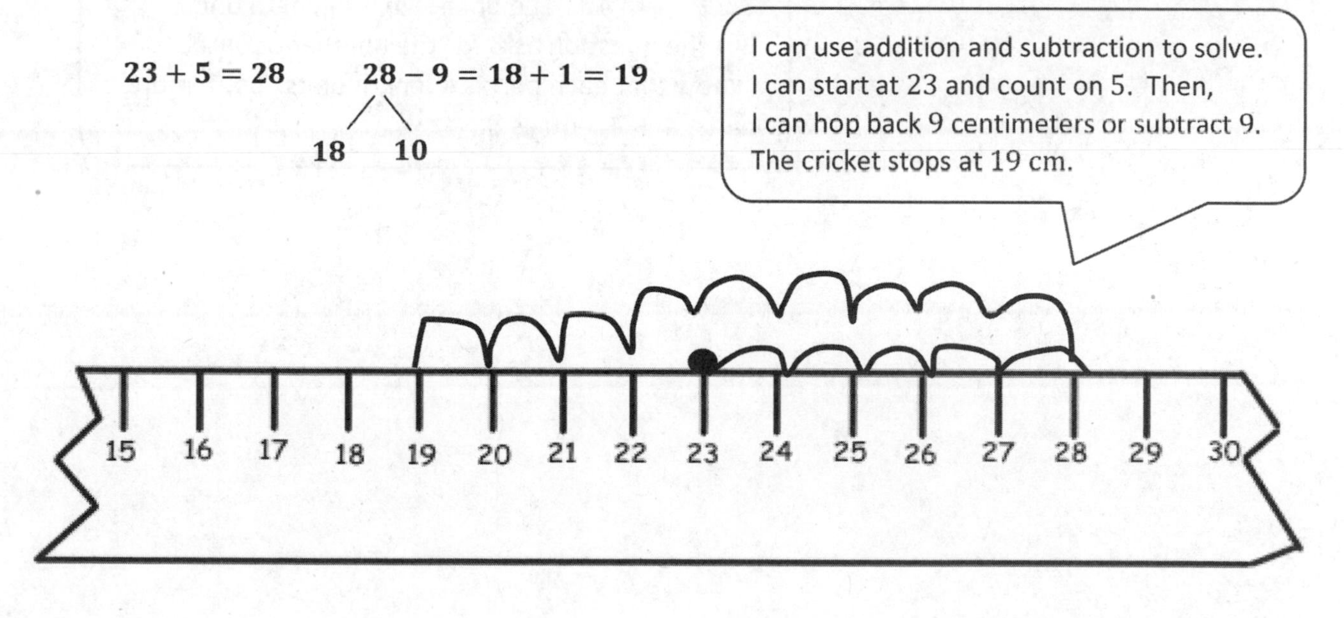

3. All of the parts of the path below are equal length units. Fill in the lengths of each side.

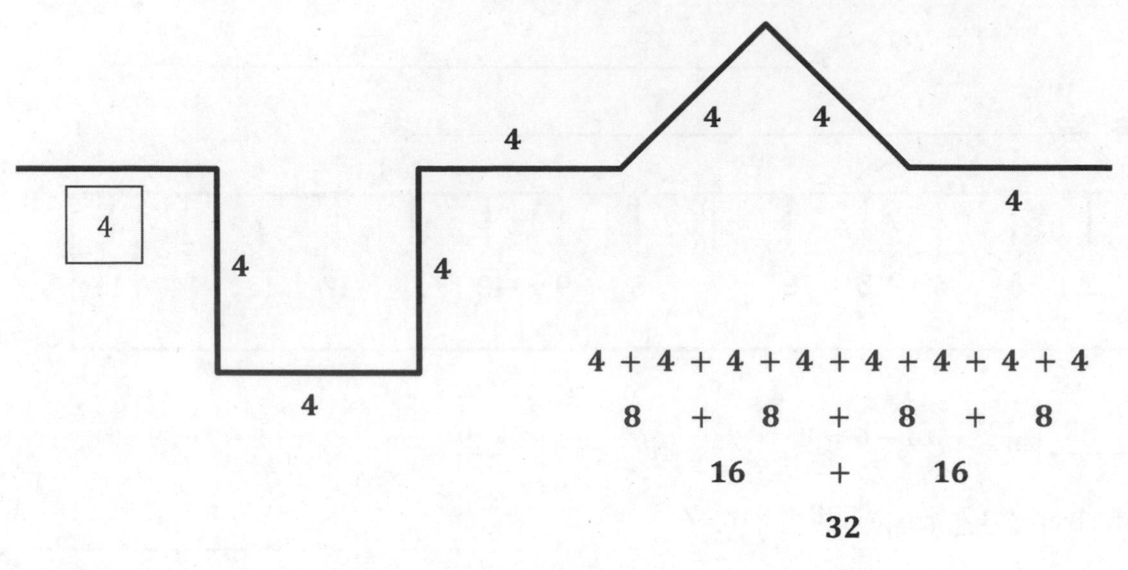

The path is __32__ length units long.

How many more parts would you need to add for the path to be 40 length units long? __2__ parts

I know that the path is 32 length units. I can think $32 + \_\_ = 40$. The unknown is 8 length units. But the question asks for the number of parts. I know that each part is 4 length units. So, 2 more parts, $4 + 4$, equals 8.

Name ______________________________ Date ______________

1.

C

D

12 13 14 15 16 17 18 19 20 21 22 23 24 25 26

a. Line C is ______ cm.

b. Line D is ______ cm.

c. Lines C and D are ______ cm.

d. Line C is ______ cm (longer/shorter) than Line D.

2. An ant walked 12 centimeters to the right on the ruler and then turned around and walked 5 centimeters to the left. His starting point is marked on the ruler. Where is the ant now? Show your work on the broken ruler.

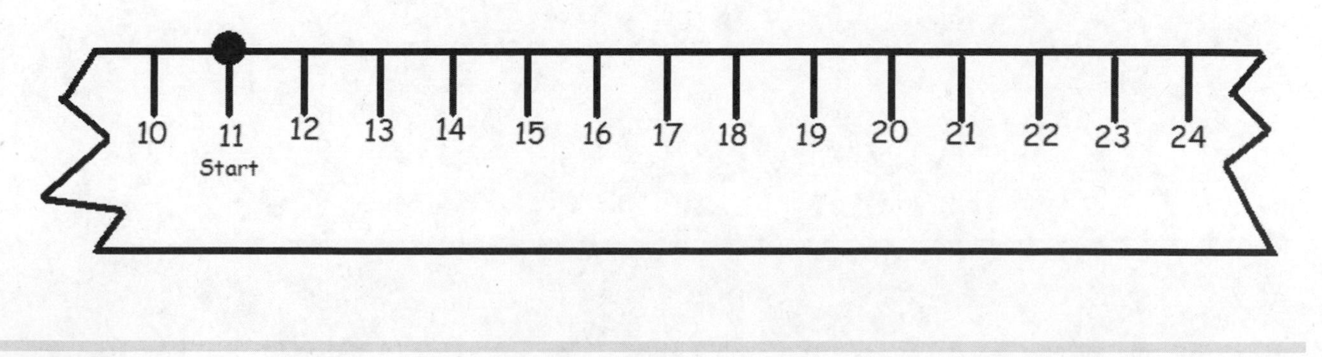

3. All of the parts of the path below are equal length units.

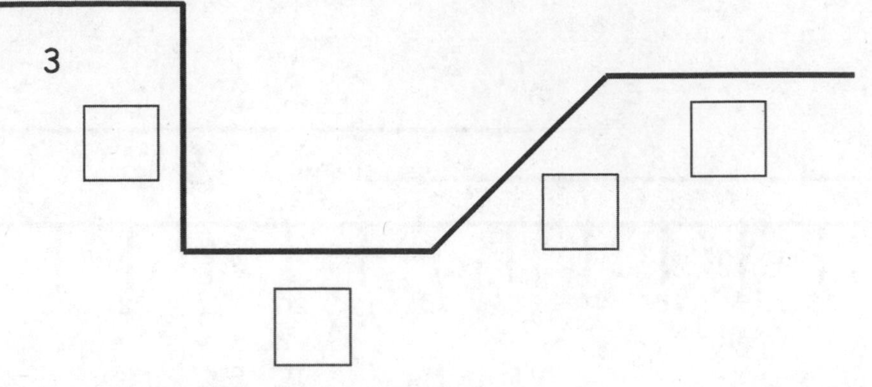

   a. Fill in the empty boxes with the lengths of each side.

   b. The path is _____ length units long.

   c. How many more parts would you need to add for the path to be 21 length units long?

      _____ parts

4. The length of a picture is 67 centimeters. The width of the picture is 40 centimeters. How many more centimeters is the length than the width?

1. Tommy completed the chart below by first estimating the measurement around three body parts and then finding the actual measurement with his meter strip.

Body Part Measured	Estimated Measurement in Centimeters	Actual Measurement in Centimeters
Neck	25 cm	31 cm
Wrist	13 cm	17 cm
Head	50 cm	57 cm

What is the difference between the longest and shortest measurements?

40 cm $57 - 17 = 40$

Draw a tape diagram comparing the measurements of Tommy's neck and wrist.

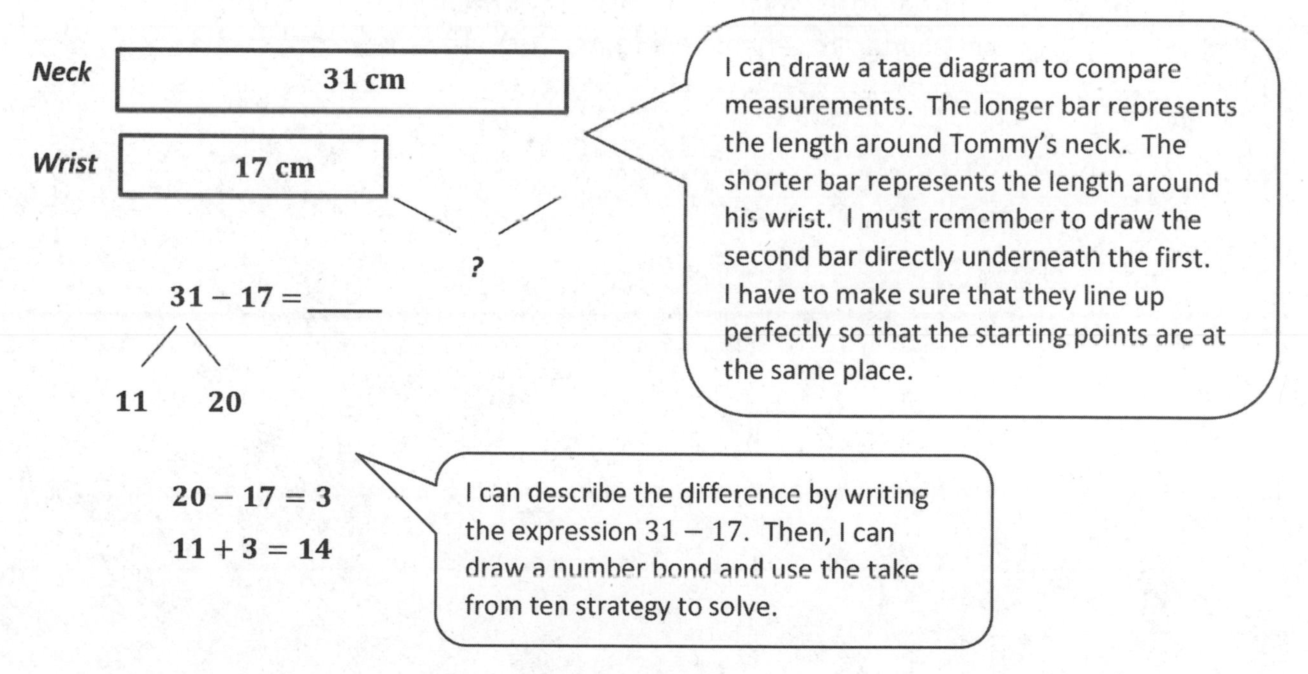

2. Measure the two paths below with your meter strip and string.

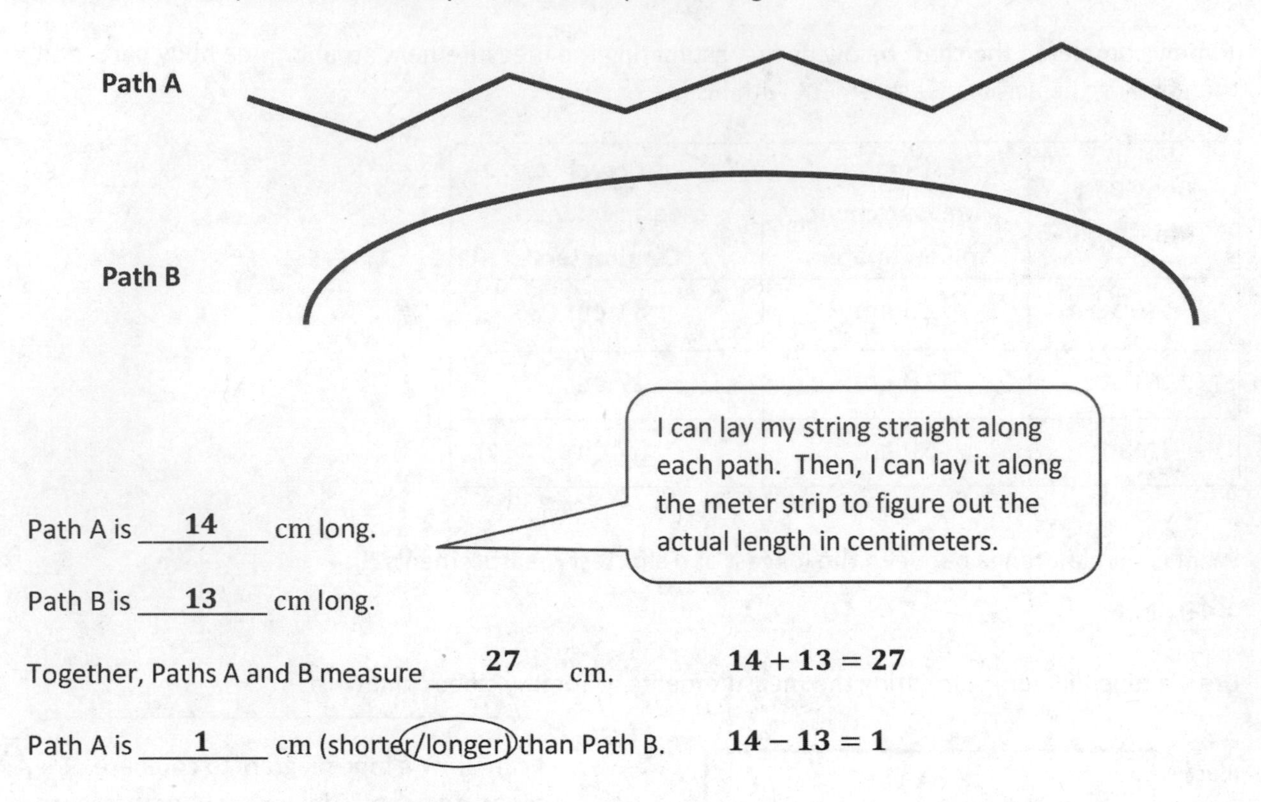

Path A is **14** cm long.

Path B is **13** cm long.

Together, Paths A and B measure **27** cm. $14 + 13 = 27$

Path A is **1** cm (shorter/longer) than Path B. $14 - 13 = 1$

Name ______________________________ Date ______________

1. Mia completed the chart by first estimating the measurement around three objects in her house and then finding the actual measurement with her meter strip.

Object Name	Estimated Measurement in Centimeters	Actual Measurement in Centimeters
Orange	40 cm	36 cm
Mini Basketball	30 cm	41 cm
Bottom of a glue bottle	10 cm	8 cm

a. What is the difference between the longest and shortest measurements?

_________ cm

b. Draw a tape diagram comparing the measurements of the orange and the bottom of the glue bottle.

c. Draw a tape diagram comparing the measurements of the basketball and the bottom of the glue bottle.

2. Measure the two paths below with your meter strip and string.

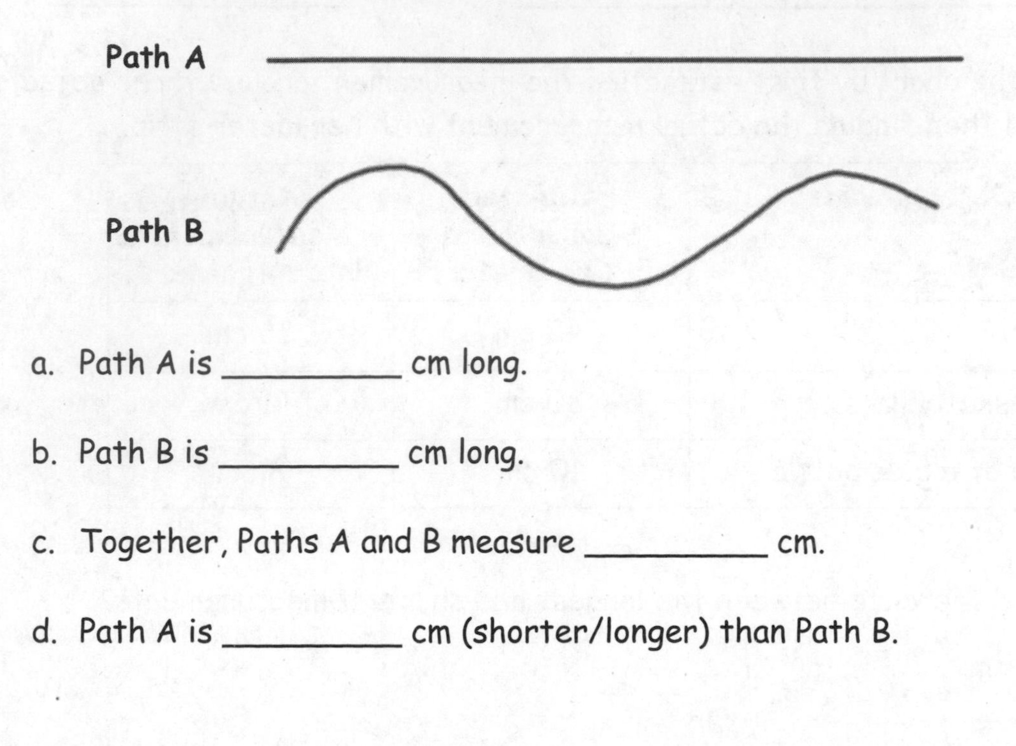

a. Path A is __________ cm long.

b. Path B is __________ cm long.

c. Together, Paths A and B measure __________ cm.

d. Path A is __________ cm (shorter/longer) than Path B.

3. Shawn and Steven had a contest to see who could jump farther. Shawn jumped 75 centimeters. Steven jumped 9 more centimeters than Shawn.

a. How far did Steven jump? __________ centimeters

b. Who won the jumping contest? __________

c. Draw a tape diagram to compare the lengths that Shawn and Steven jump.

Use the Read-Draw-Write (RDW) process to solve. Draw a tape diagram for each step.

Jesse's tower of blocks is 30 cm tall. Sarah's tower is 9 cm shorter than Jesse's tower. What is the total height of both towers?

Step 1: Find the height of Sarah's tower.

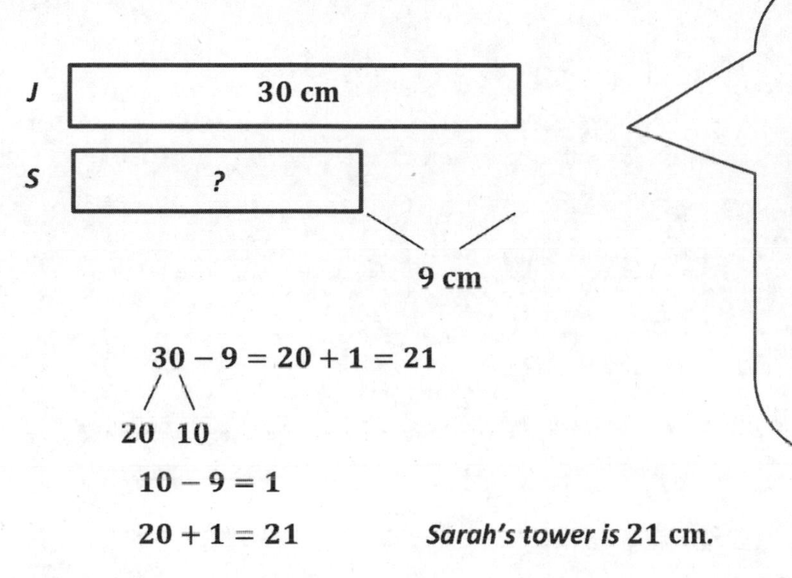

I can draw a tape diagram to compare Jesse and Sarah's towers. I don't know how tall Sarah's tower is, so I can label it with a question mark. But, I know that Sarah's tower is shorter, so I can draw arms and label the difference with 9 cm. I can use subtraction and the take from ten strategy to find the missing part, so $30 - 9 = 21$.

Step 2: Find the total height of both towers.

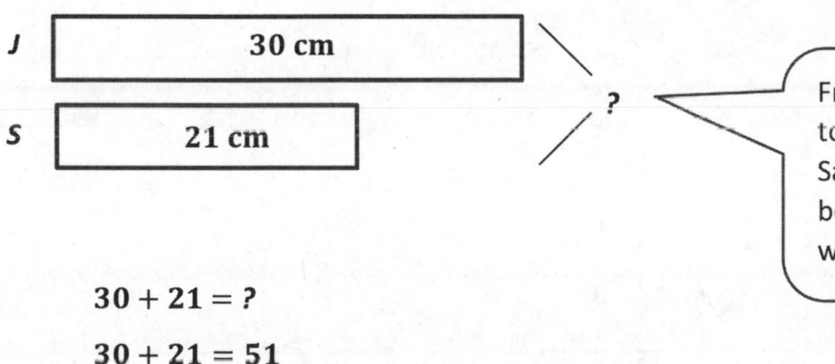

From Step 1, I know that Sarah's tower is 21 cm. Now, I can label Sarah's bar with 21 cm. I can add both parts together to find the whole. $30 \text{ cm} + 21 \text{ cm} = 51 \text{ cm}$.

***The total height of both towers is* 51 cm.**

Name ______________________________ Date ______________

Use the RDW process to solve. Draw a tape diagram for each step. Problem 1 has been started for you.

1. There is 29 cm of green ribbon. A blue ribbon is 9 cm shorter than the green ribbon. How long is the blue ribbon?

   Step 1: Find the length of blue ribbon.

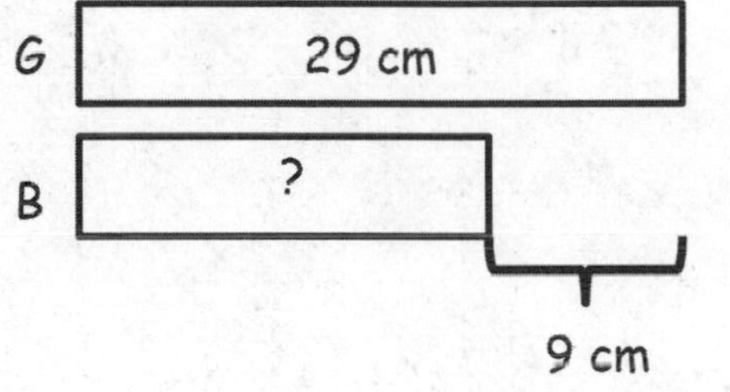

   Step 2: Find the length of both the blue and green ribbons.

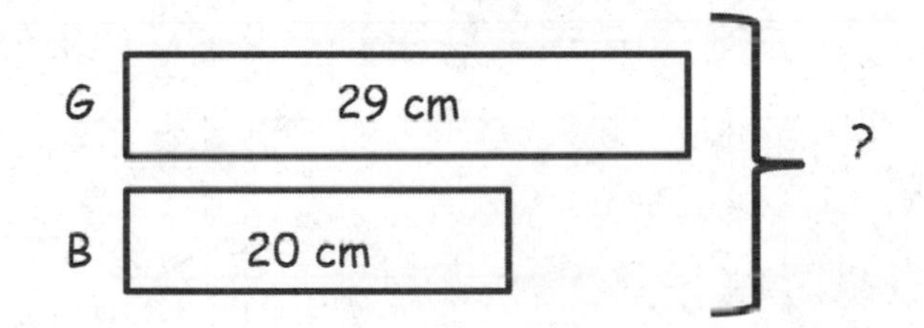

2. Joanna and Lisa drew lines. Joanna's line is 41 cm long. Lisa's line is 19 cm longer than Joanna's. How long are Joanna's and Lisa's lines?

   Step 1: Find the length of Lisa's line.

   Step 2: Find the total length of their lines.

Grade 2
Module 3

1. Fill in the missing part.

   a. 3 ones + __7__ ones = 10 ones

   b. 3 + __7__ = 10

   c. 3 tens + __7__ tens = 1 hundred

   d. 30 + __70__ = 100

> I know 3 facts that can help me solve all these problems:
> $3 + 7 = 10$
> $10 \text{ ones} = 1 \text{ ten}$
> $10 \text{ tens} = 1 \text{ hundred}$

2. Rewrite in order from largest to smallest units.

4 tens	Largest ___*2 hundreds*___
2 hundreds	___*4 tens*___
9 ones	Smallest ___*9 ones*___

> I know that 2 hundreds equal 200, 4 tens equal 40, and 9 ones equal 9.

3. Count each group. What is the total number of sticks in each group?

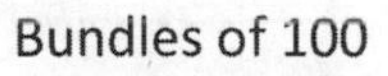

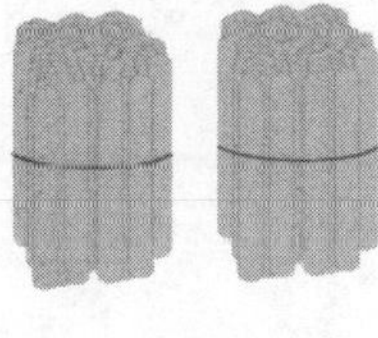

__200__

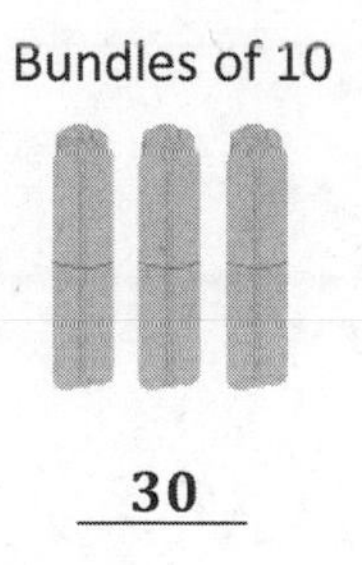

__30__

Ones

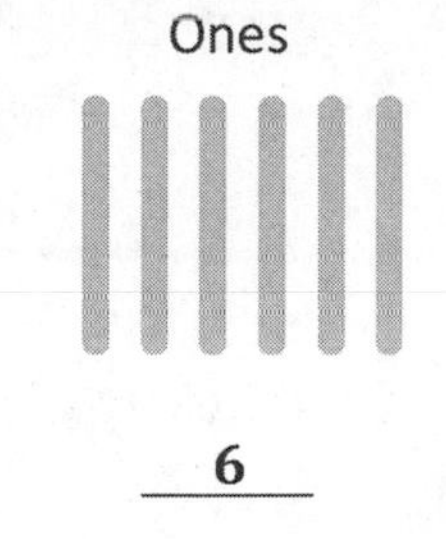

__6__

What is the total number of sticks? __236__

4. Draw and solve.

   Moses has 100 stickers. Jared has 80 stickers. Jared wants to have the same number of stickers as Moses. How many more stickers does Jared need?

I can start at 80 and count on by 10's until I reach 100.

I can draw bundles of 10 to help me keep count: 90, 100.

Jared needs __20__ more stickers

I counted 2 more tens. That's 20.

Name ______________________ Date ____________

1. 2 ones + ____ ones = 10

   2 + ____ = 10

2. 6 tens + ____ tens = 1 hundred

   60 + ____ = 100

3. Rewrite in order from largest to smallest units.

6 tens	Largest	______________________
3 hundreds		______________________
8 ones	Smallest	______________________

4. Count each group. What is the total number of sticks in each group?

Bundles of 100 *Bundles of 10* *Ones*

__________ __________ __________

What is the total number of sticks? ______

5. Draw and solve.

Moses has 100 stickers. Jared has 60 stickers. Jared wants to have the same number of stickers as Moses. How many more stickers does Jared need?

Jared needs _____ more stickers.

1. These are bundles with 10 sticks in each.

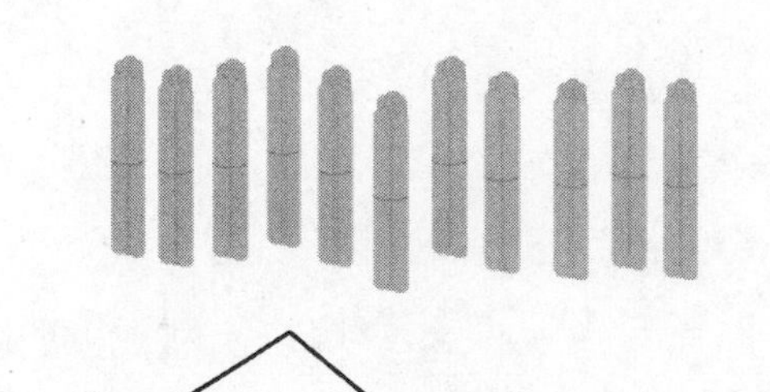

   a. How many tens are there? **11**

   b. How many hundreds? **1**

   c. How many sticks in all? **110**

> I count 11 tens. I know that 10 tens equal 1 hundred. I can skip-count by tens to see that there are 110 sticks in all.

2. Dean did some counting. Look at his work. Explain why you think Dean counted this way.

   128, 129, 130, 140, 150, 160, 170, 180, 181, 182, 183

> Benchmark numbers allow us to skip-count, which is faster than counting by ones. So Dean counted by ones to get to the closest benchmark number, 130. Then, he skip-counted by tens up to 180. Next, he counted by ones to reach 183.

3. Show a way to count from 76 to 140 using tens and ones. Explain why you chose to count this way.

   76, 77, 78, 79, 80, 90, 100, 110, 120, 130, 140

> I counted by ones to get to the nearest benchmark number after 76, which is 80. Then it was easy to skip-count by tens up to 140.

Name ______________________________ Date ________________

1. How many in all?

☆☆ ☆☆ ☆☆ ☆☆

☆☆ ☆☆ ☆☆ ☆☆

☆☆ ☆☆ ☆☆ ☆☆

☆☆ ☆☆ ☆☆ ☆☆

☆☆ ☆☆ ☆☆ ☆☆

_____ ones = ______ tens

_____ stars in all.

2. These are bundles with 10 sticks in each.

a. How many tens are there? _______

b. How many hundreds? _______

c. How many sticks in all? _______

3. Sally did some counting. Look at her work. Explain why you think Sally counted this way.

177, 178, 179, 180, 190, 200, 210, 211, 212, 213, 214

4. Show a way to count from 68 to 130 using tens and ones. Explain why you chose to count this way.

5. Draw and solve.

   In her classroom, Sally made 17 bundles of 10 straws. How many straws did she bundle in all?

I count by ones to reach 70. I count by tens to reach 100. I count by hundreds to reach 400, and then I count by tens to get to 420.

1. Fill in the blanks to reach the benchmark numbers.

66, **67**, **68**, **69**, 70, **80**, **90**, 100, **200**, **300**, 400, **410**, 420

Benchmark numbers make it quicker and easier to count to large numbers!

2. These are ones, tens, and hundreds. How many sticks are there in all?

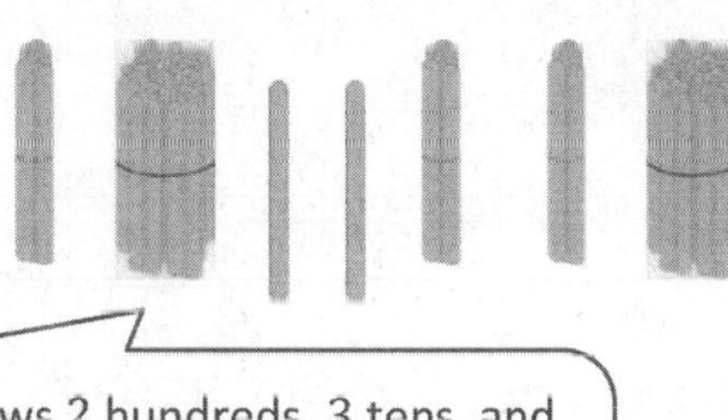

I know that the order of these drawn units doesn't matter, but it's easiest to start with the highest value, the hundreds.

This shows 2 hundreds, 3 tens, and 2 ones. I can count like this: 100, 200, 210, 220, 230, 231, 232. So there are 232 sticks in all.

There are **232** sticks in all.

3. Show a way to count from 457 to 700 using ones, tens, and hundreds.

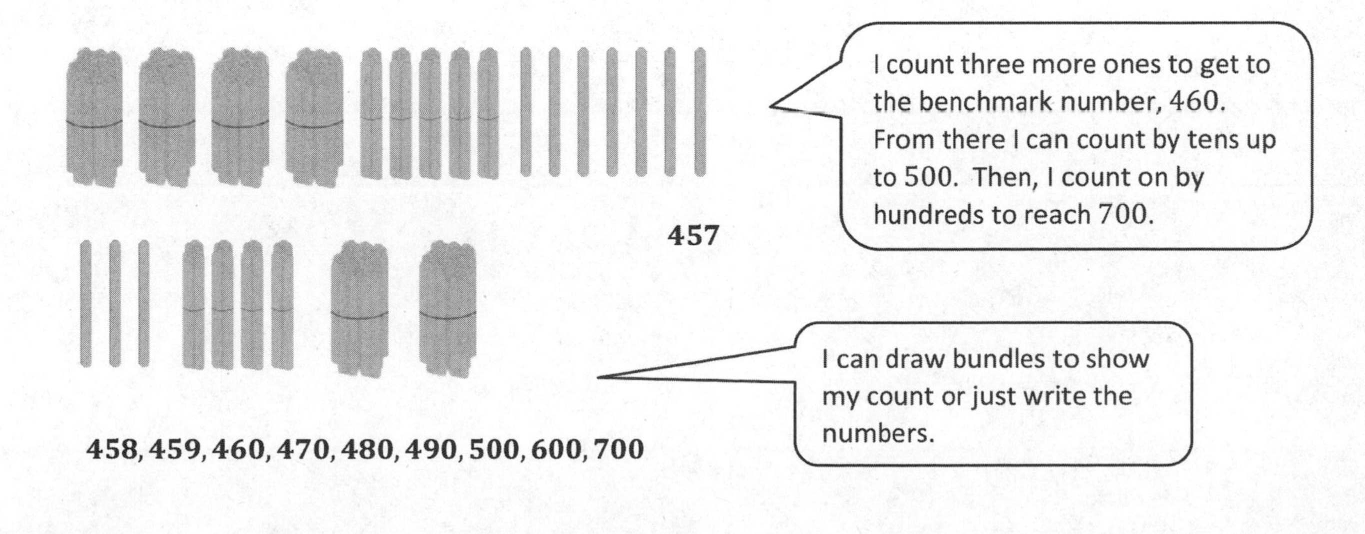

I count three more ones to get to the benchmark number, 460. From there I can count by tens up to 500. Then, I count on by hundreds to reach 700.

I can draw bundles to show my count or just write the numbers.

Name ______________________________ Date ______________

1. Fill in the blanks to reach the benchmark numbers.

   a. 14, _____, _____, _____, _____, _____, 20, _____, _____, 50

   b. 73, _____, _____, _____, _____, _____, _____, 80, _____, 100, _____, 300, _____, 320

   c. 65, _____, _____, _____, _____, 70, _____, _____, 100

   d. 30, _____, _____, _____, _____, _____, _____, 100, _____, _____, 400

2. These are ones, tens, and hundreds. How many sticks are there in all?

   There are __________ sticks in all.

3. Show a way to count from 668 to 900 using ones, tens, and hundreds.

4. Sally bundled her sticks in hundreds, tens, and ones.

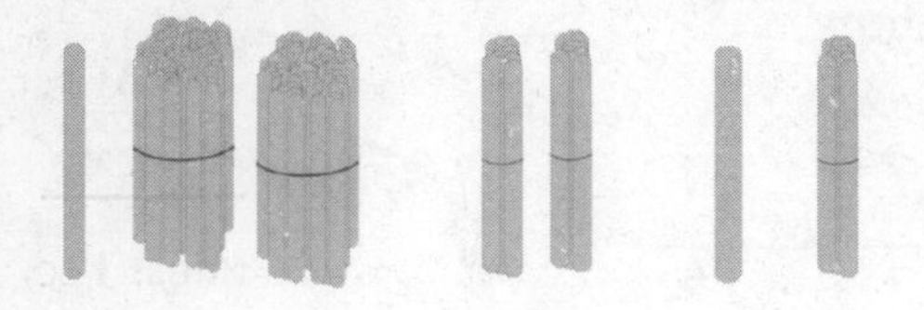

a. How many sticks does Sally have? ______________________

b. Draw 3 more hundreds and 3 more tens. Count and write how many sticks Sally has now.

1. Pilar used the place value chart to count bundles. How many sticks does she have in all?

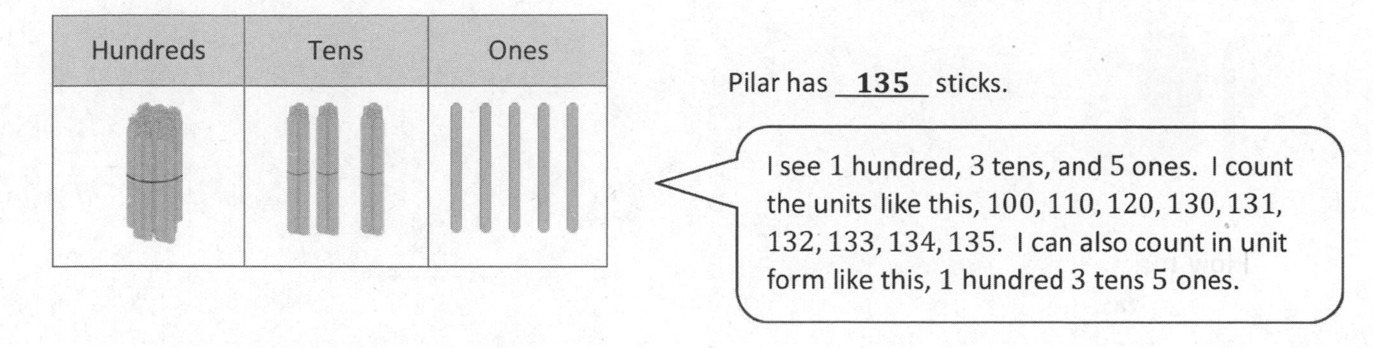

Pilar has __135__ sticks.

2. These are tens. If you put them together, which unit will you make?

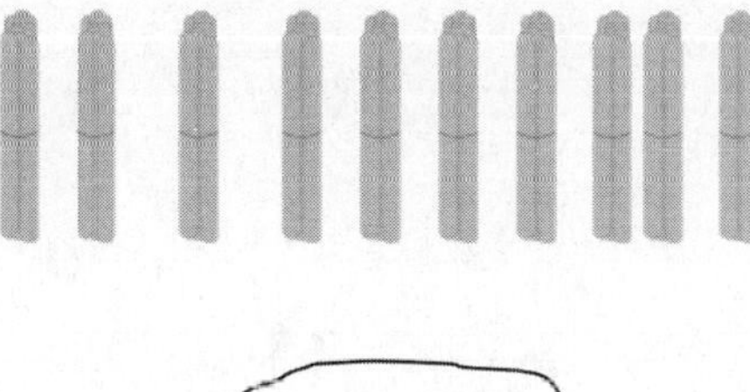

I can skip-count by ten to see that 10 tens equal 1 hundred.
10, 20, 30, 40, 50, 60, 70, 80, 90, 100.
I can bundle it to show 100.

a. one (b. hundred) c. thousand d. ten

3. Imagine 467 on the place value chart. How many ones, tens, and hundreds are in each place?

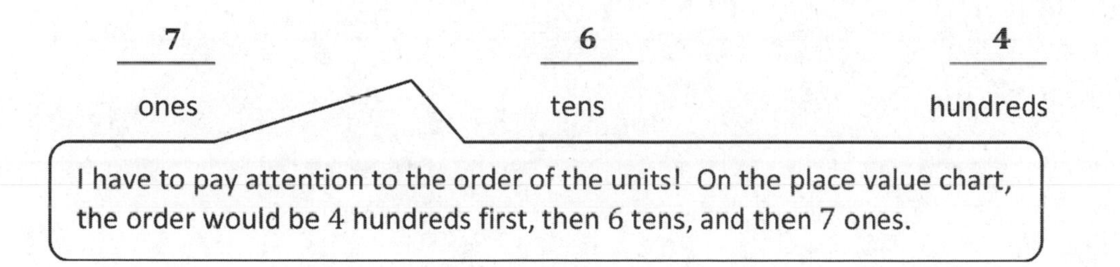

4. Show a way to count from 160 to 530 using tens and hundreds. Circle at least one benchmark number.

160, 170, 180, 190, (200), 300, 400, 500, 510, 520, 530

I skip-count by tens to reach 200. After that, I can count on by hundreds. At 500, I count by tens to reach 530.

Name ______________________________ Date ________________

1. Marcos used the place value chart to count bundles. How many sticks does Marcos have in all?

Hundreds	Tens	Ones

Marcos has _____________ sticks.

2. Write the number:

Hundreds	Tens	Ones

3. These are hundreds. If you put them together, which unit will you make?

a. one b. hundred c. thousand d. ten

4. Imagine 585 on the place value chart. How many ones, tens, and hundreds are in each place?

    ______________ ones   ______________ tens   ______________ hundreds

5. Fill in the blanks to make a true number sentence.

    12 ones = _____ ten _____ ones

6. Show a way to count from 170 to 410 using tens and hundreds. Circle at least 1 benchmark number.

7. Mrs. Sullivan's students are collecting cans for recycling. Frederick collected 20 cans, Donielle collected *9 cans,* and Mina and Charlie each collected 100 cans. How many cans did the students collect in all?

1. What is the value of the 5 in | 8 | 5 | 9 | ? __50__

800
50
9

I can picture how this number looks when shown with Hide Zero cards. The digit 5 is in the tens place. I know the value of 5 tens is 50.

2. Make a number bond to show the hundreds, tens, and ones in the number. Then, write the number in unit form.

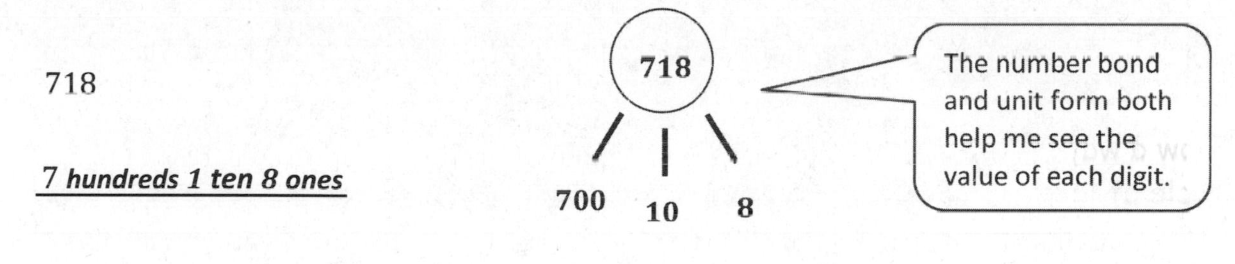

3. Draw a line to match unit form with number form.

a. 4 hundreds 1 ten = 41

b. 4 tens 1 one = 410

c. 4 hundreds 1 one = 401

I can visualize the numbers on the place value chart to help me match the unit form to number form.

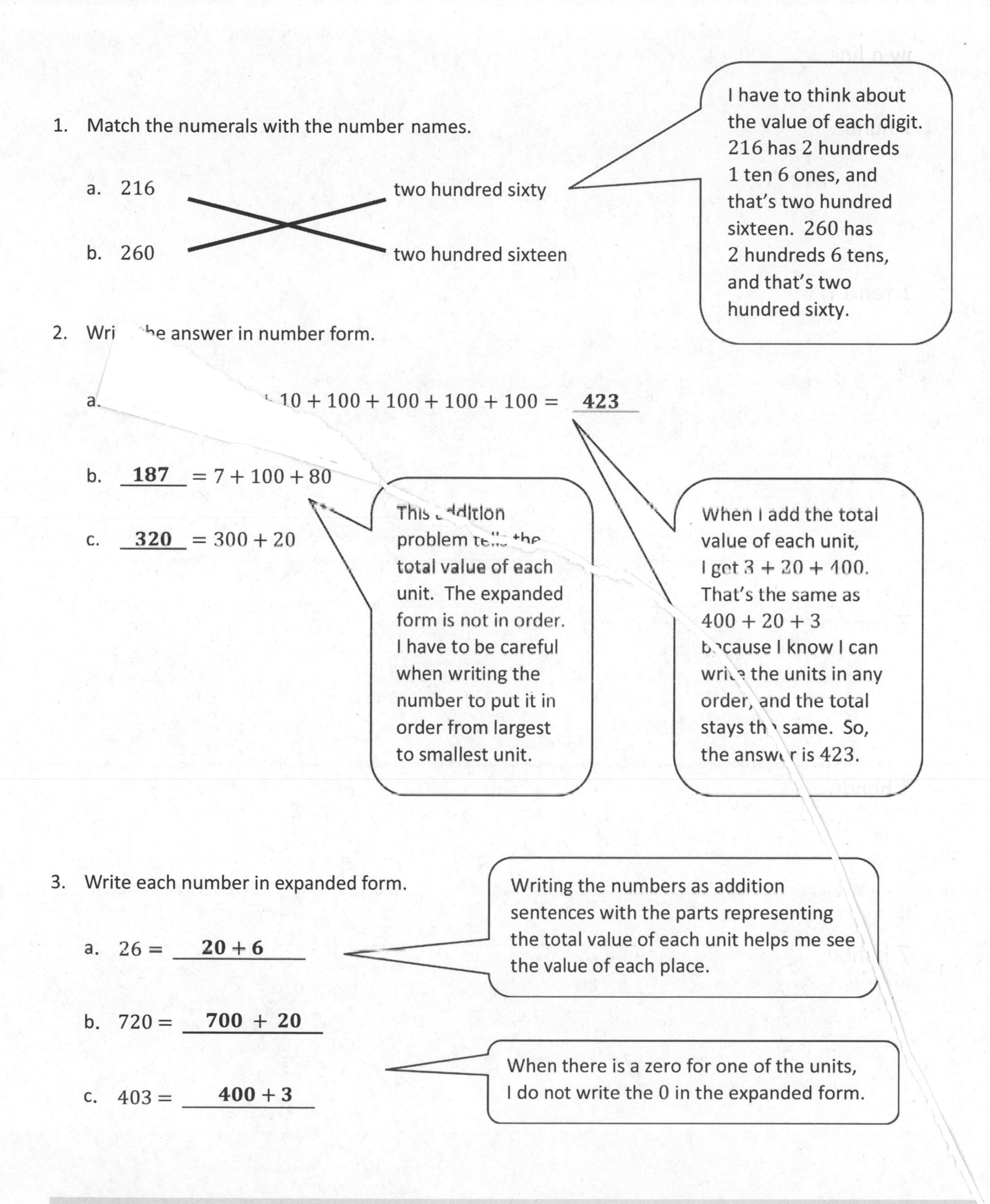

1. Match the numerals with the number names.

   a. 216 — two hundred sixteen

   b. 260 — two hundred sixty

2. Wri he answer in number form.

   a. 10 + 100 + 100 + 100 + 100 = **423**

   b. **187** = 7 + 100 + 80

   c. **320** = 300 + 20

3. Write each number in expanded form.

   a. 26 = **20 + 6**

   b. 720 = **700 + 20**

   c. 403 = **400 + 3**

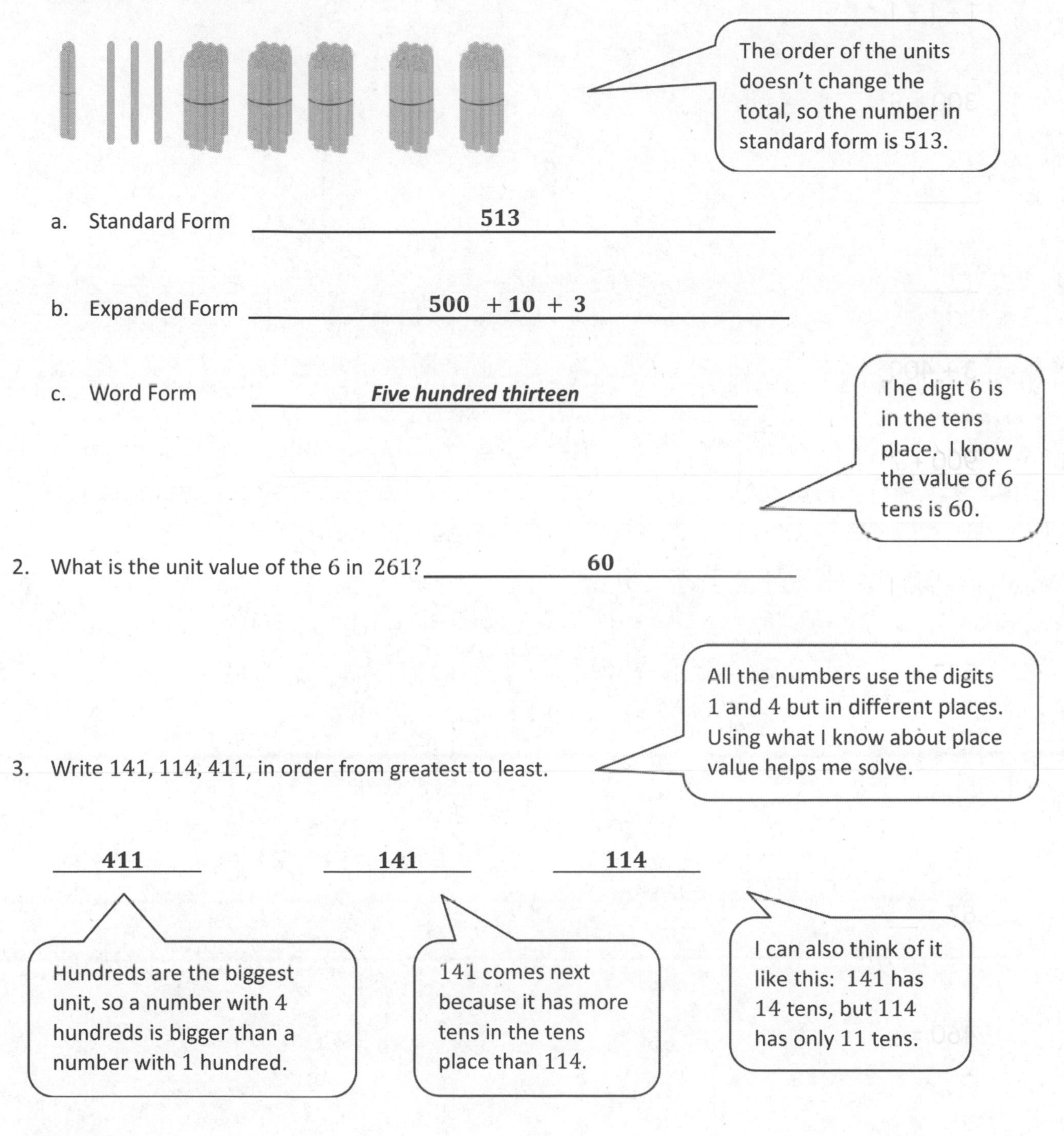

1. These are bundles of hundreds, tens, and ones. Write the standard form, expanded form, and word form for each number shown.

a. Standard Form **513**

b. Expanded Form **500 + 10 + 3**

c. Word Form ***Five hundred thirteen***

2. What is the unit value of the 6 in 261? **60**

3. Write 141, 114, 411, in order from greatest to least.

411 **141** **114**

1. Write the total value of the money.

2. Fill in the bills with \$100, \$10, or \$1 to show the amount.

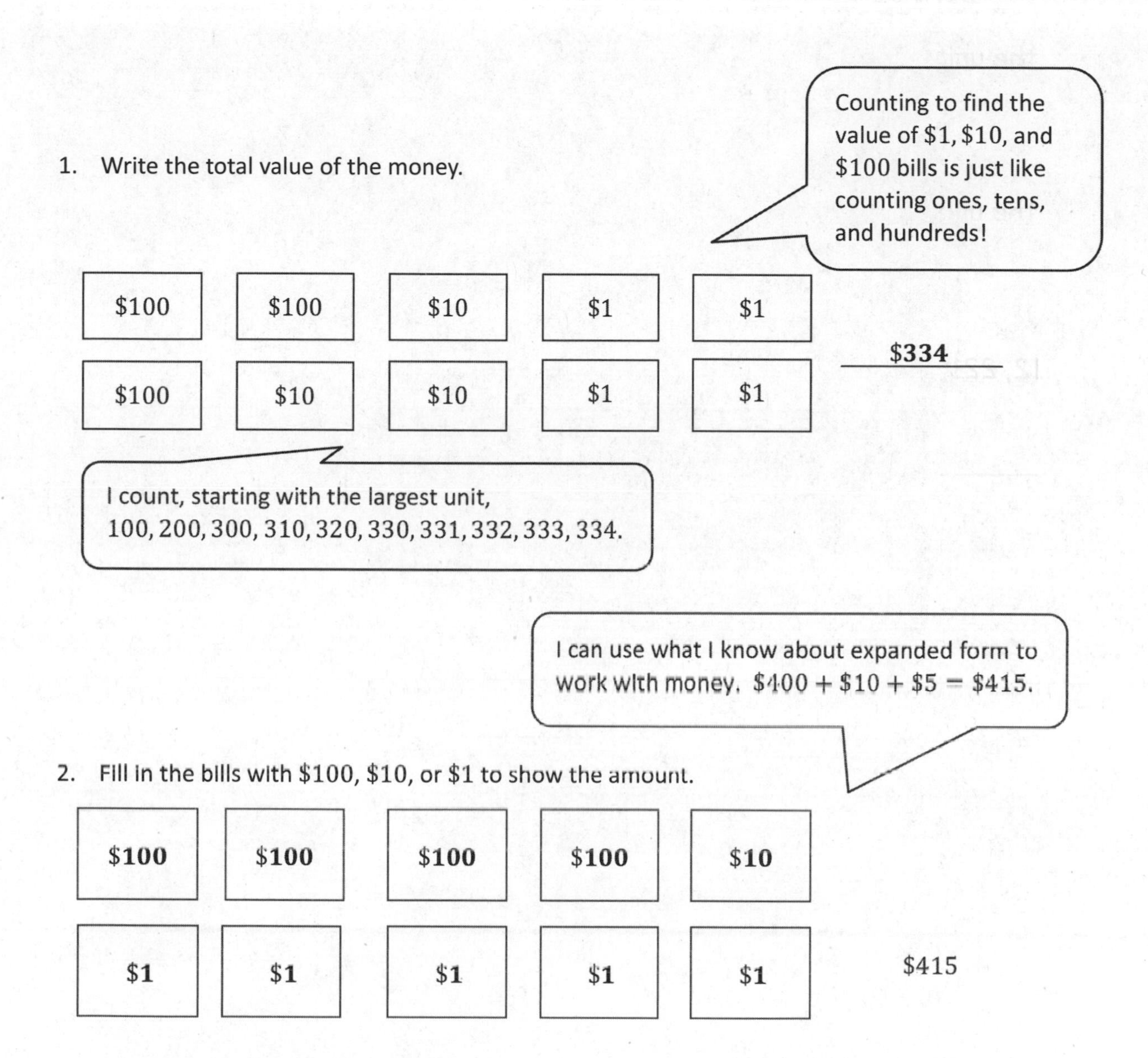

3. Draw and solve.

Jill has 5 ten-dollar bills and 3 one-dollar bills. Ben has 2 fewer ten-dollar bills and 1 fewer one-dollar bill than Jill. What is the value of Ben's money?

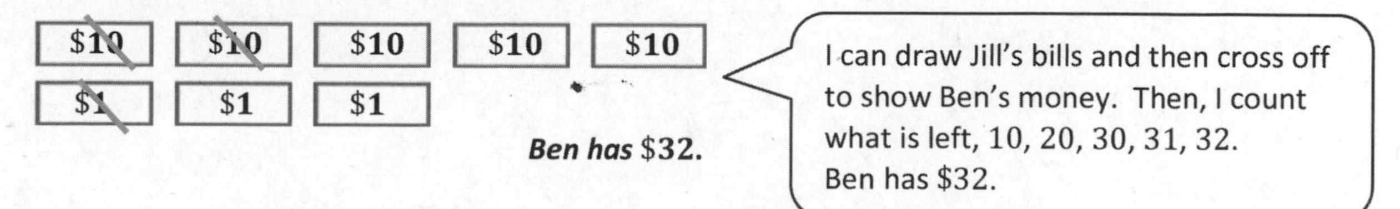

Name ______________________ Date __________

1. Write the total value of the money.

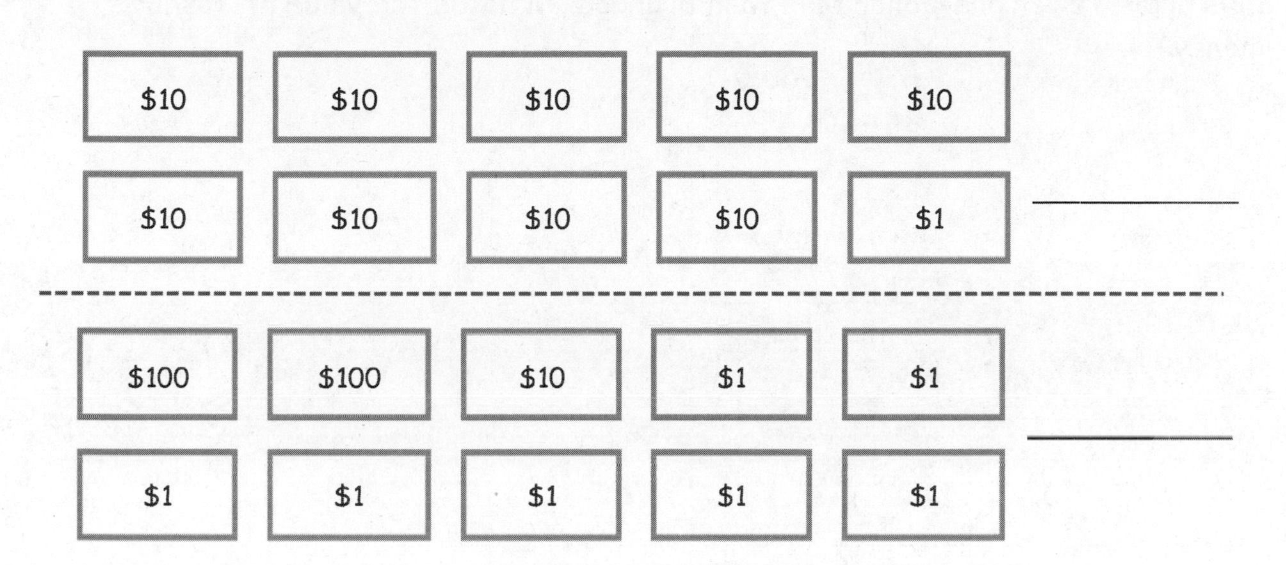

2. Fill in the bills with $100, $10, or $1 to show the amount.

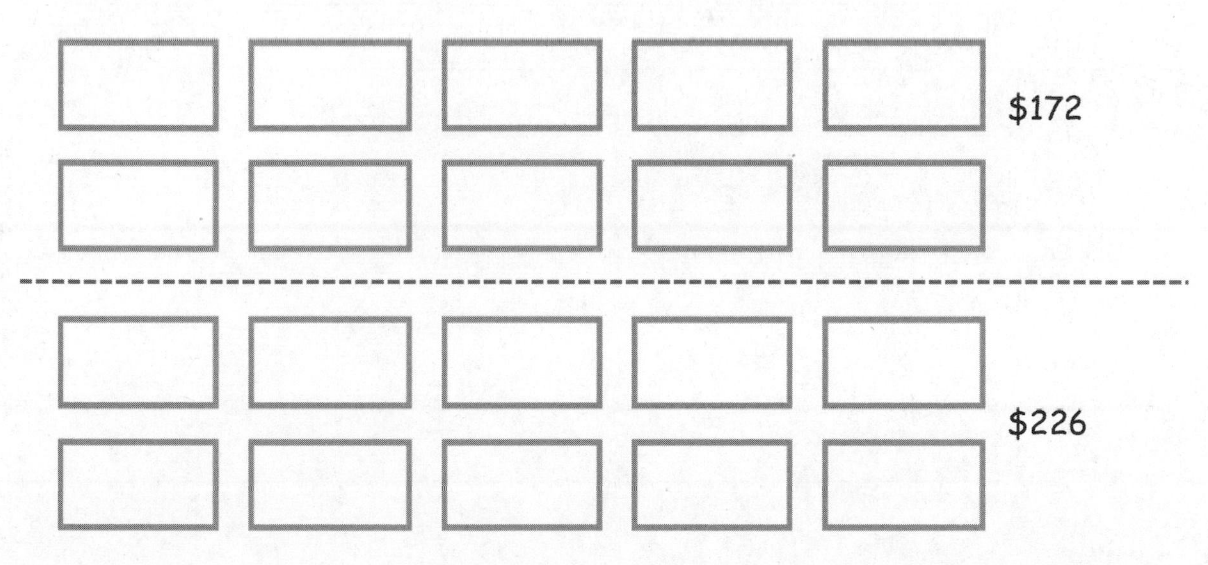

3. Draw and solve.

   Brandon has 7 ten-dollar bills and 8 one-dollar bills. Joshua has 3 fewer ten-dollar bills and 4 fewer one-dollar bills than Brandon. What is the value of Joshua's money?

1. Show one way to count from $67 to $317.

67, 77, 87 97, 107, 117, 217, 317

Counting money is just like counting with numerals, so I can leave off the dollar signs and just skip-count by tens to get to 117. Then, I skip-count by hundreds to get to 317.

2. Use each number line to show a different way to count from $280 to $523.

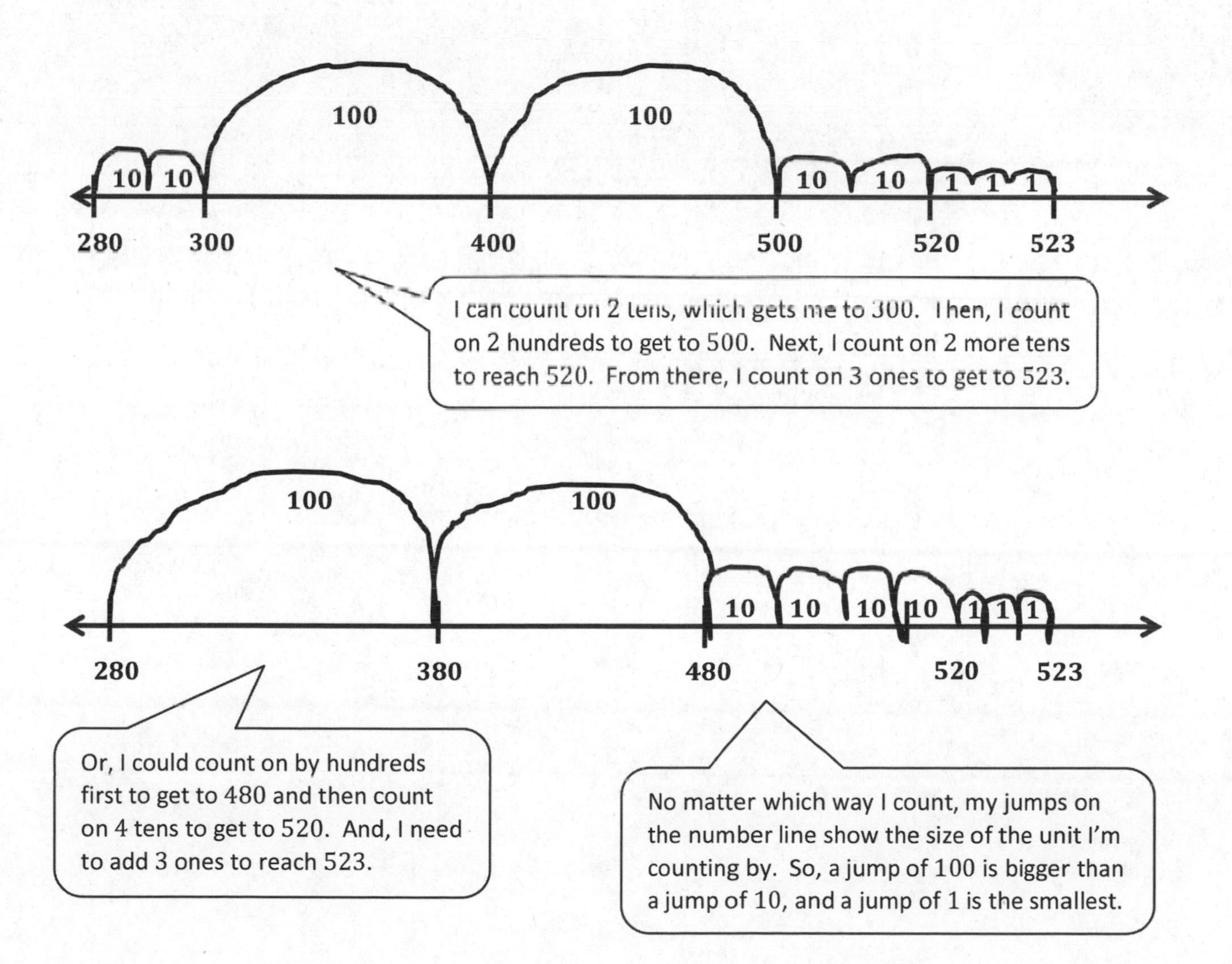

Name ______________________________ Date ______________

1. Write the total amount of money shown in each group.

a.

$100	$100
$100	$100
$100	$100
$100	$100
$100	$100

b.

$10	$10
$10	$10
$10	$10
$10	$10
$10	$10

c.

$1	$1
$1	$1
$1	$1
$1	$1
$1	$1

d.

$10	$100
$10	$100
$10	$100
$100	$1
$100	$1

__________ __________ __________ __________

2. Show one way to count from $82 to $512.

3. Use each number line to show a different way to count from $580 to $994.

4. Draw and solve.

   Julia wants a bike that costs $75. She needs to save $25 more to have enough money to buy it. How much money does Julia already have?

   Julia already has $___________.

How many $10 bills are equal to $500?

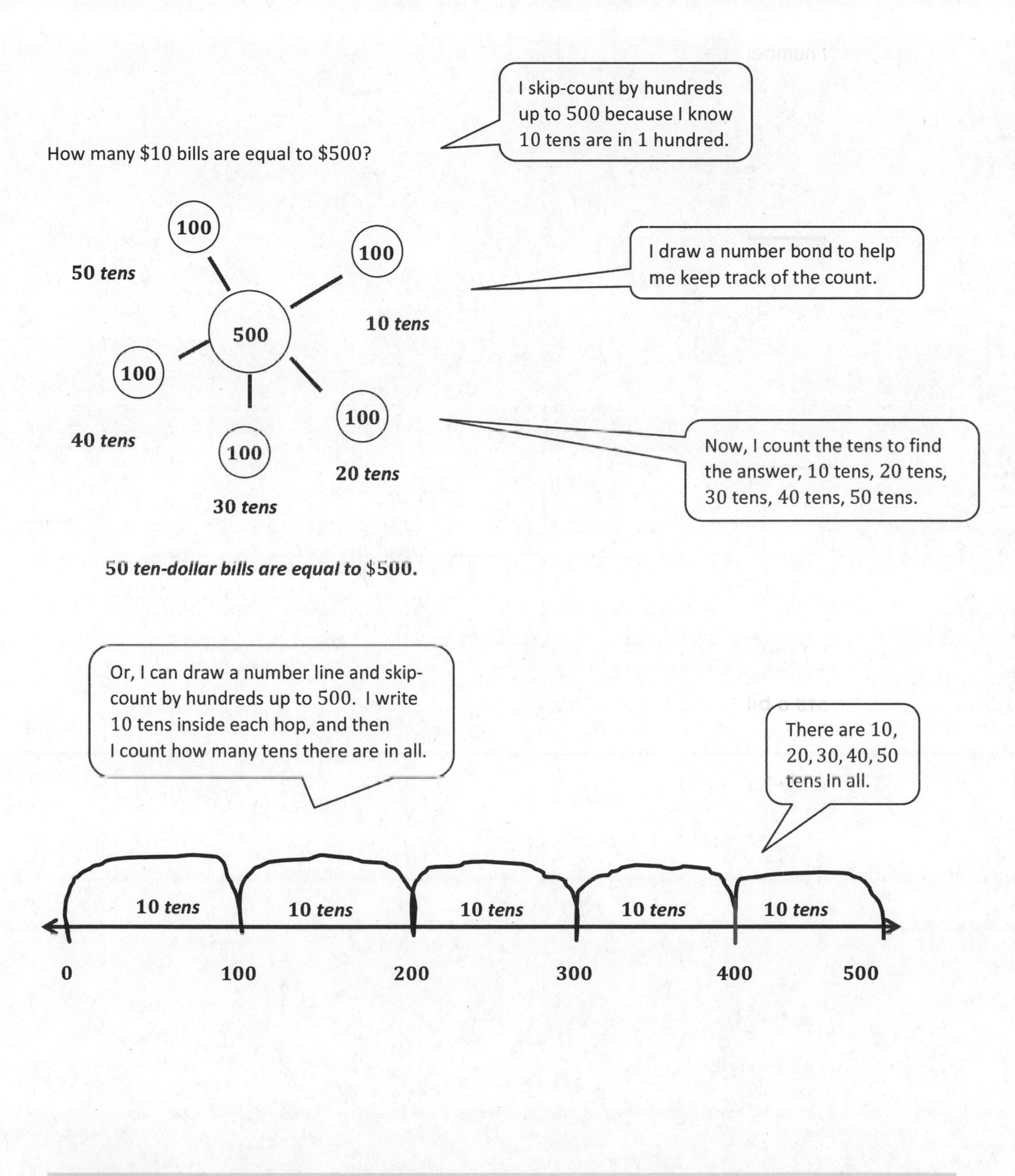

Name ______________________________ Date ______________

Jerry wonders, "How many $10 bills are equal to a $1,000 bill?"

Think about the strategies your friends used to answer Jerry's question. Answer the problem again using a different strategy than the one you used with your partner and for the Exit Ticket. Explain your solution using words, pictures, or numbers. Remember to write your answer as a statement.

Students use place value disks to model the value of each digit in a given number. A template has been provided to help students complete the homework assignment.

Model the following numbers for your parent using the fewest disks possible. Whisper the numbers in standard form and unit form.

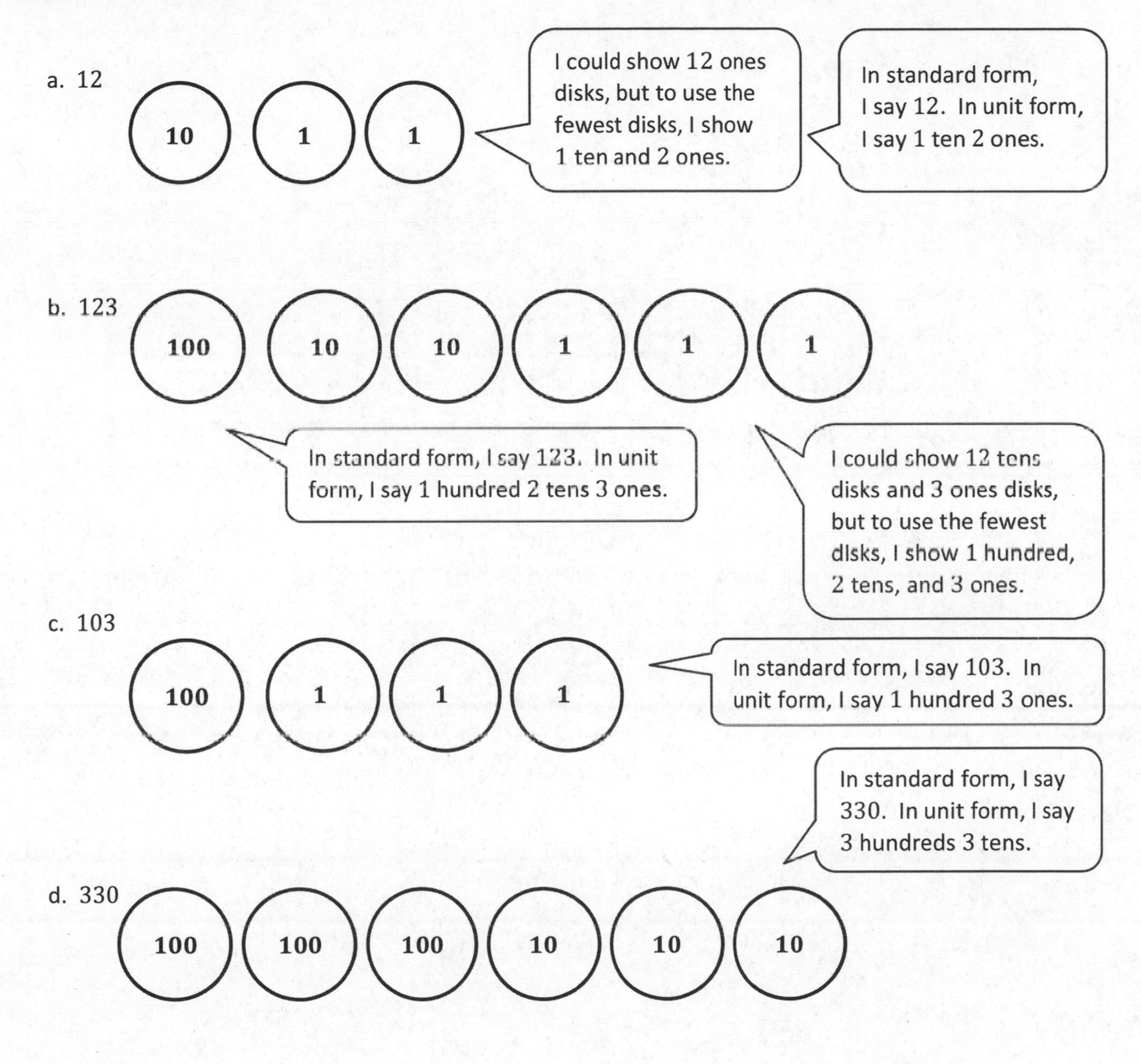

Name ______________________________ Date ______________

1. Model the following numbers for your parent using the fewest disks possible. Whisper the numbers in standard form and unit form (1 hundred 3 tens 4 ones).

   a. 15

   b. 152

   c. 102

   d. 290

   e. 300

2. Model the following numbers using the fewest place value disks possible. Whisper the numbers in standard form and unit form.

   a. 42

   b. 420

   c. 320

   d. 402

   e. 442

   f. 53

   g. 530

   h. 520

   i. 503

   j. 55

unlabeled hundreds place value chart; from Lesson 8

Students complete this chart as they work with place value disks.

Count from 582 to 700 using place value disks. Change for a larger unit when necessary.

When you counted from 582 to 700:

Did you make a larger unit at...	Yes, I changed to make:	No, I need ________
1. 590?	(1 ten) 1 hundred	___ ones. ___ tens.
2. 600?	1 ten (1 hundred)	___ ones. ___ tens.
3. 618?	1 ten 1 hundred	2 ones. ___ tens.
4. 640?	(1 ten) 1 hundred	___ ones. ___ tens.
5. 652?	1 ten 1 hundred	8 ones. ___ tens.
6. 700?	1 ten (1 hundred)	___ ones. ___ tens.

When I add 8 ones to 582, I make the next ten. Now I'm at 590.

Counting on from 590, when I add 10 more ones, I make a ten, which also means I make a new hundred, 600.

I need to add 2 more ones to make a new ten and reach 620.

I make a new ten when I reach 630, and again when I reach 640.

I need to add 8 more ones to make a new ten and reach 660.

Counting on from 690, when I add 10 more ones, I make a ten, which also means I make a new hundred, 700.

Name ______________________________ Date ______________

Count by ones from **368 to 500**. Change for a larger unit when necessary.

When you counted from **368 to 500**:

Did you make a larger unit at...	Yes, I changed to make:	No I need ______
1. 377?	1 ten 1 hundred	____ ones. ____ tens.
2. 392?	1 ten 1 hundred	____ ones. ____ tens.
3. 400?	1 ten 1 hundred	____ ones. ____ tens.
4. 418?	1 ten 1 hundred	____ ones. ____ tens.
5. 463?	1 ten 1 hundred	____ ones. ____ tens.
6. 470?	1 ten 1 hundred	____ ones. ____ tens.

Draw place value disks to show the numbers.

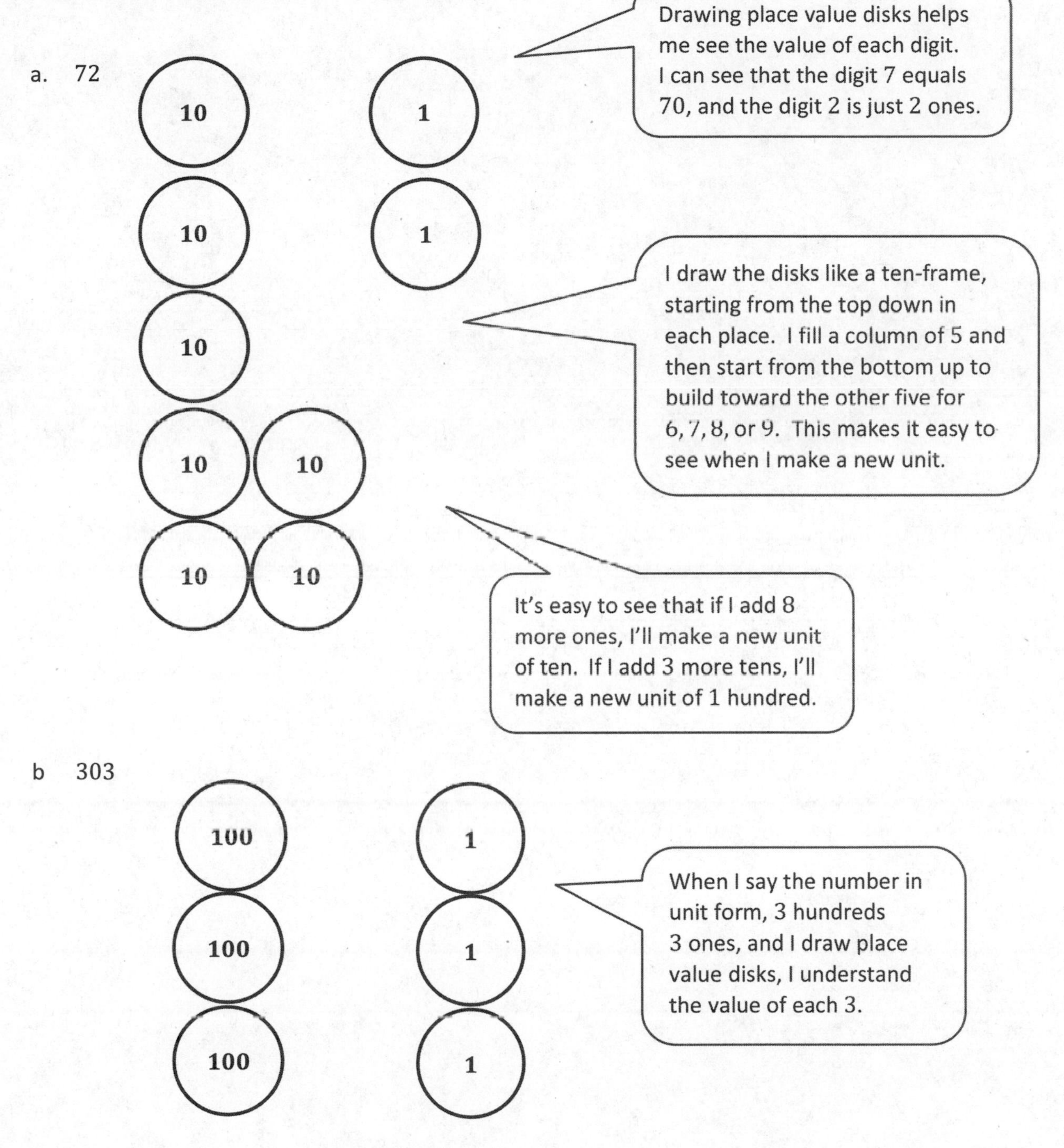

Name ______________________________ Date ______________

Draw place value disks to show the numbers.

1. 43

2. 430

3. 270

4. 720

5. 702

6. 936

When you have finished, use your whisper voice to read each number out loud in both unit and word form. How much does each number need to change for a ten? For 1 hundred?

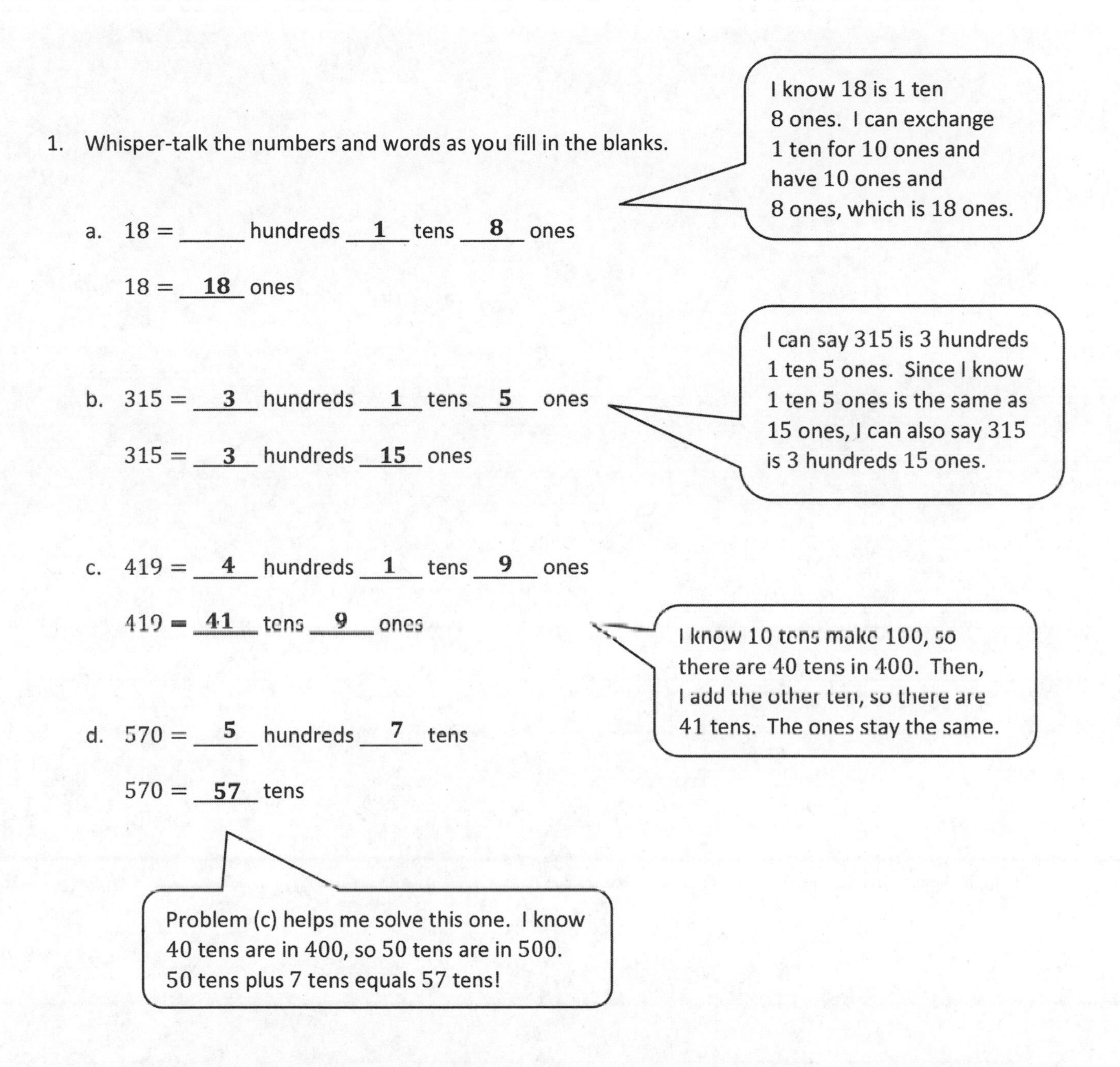

1. Whisper-talk the numbers and words as you fill in the blanks.

   a. 18 = ______ hundreds __1__ tens __8__ ones

   18 = __18__ ones

   b. 315 = __3__ hundreds __1__ tens __5__ ones

   315 = __3__ hundreds __15__ ones

   c. 419 = __4__ hundreds __1__ tens __9__ ones

   419 = __41__ tens __9__ ones

   d. 570 = __5__ hundreds __7__ tens

   570 = __57__ tens

2. Write down how you can skip-count by ten from 420 to 310. You might use place value disks, number lines, bundles, or numbers.

420, 410, 400, 390, 380, 370, 360, 350, 340, 330, 320, 310

Easy! I can just count back by ten!

Name ______________________________ Date ______________

1. Whisper-talk the numbers and words as you fill in the blanks.

a. 16 = ______ tens ______ ones

16 = ______ ones

b. 217 = ______ hundreds ______ tens ______ ones

217 = ______ hundreds ______ ones

c. 320 = ______ hundreds ______ tens ______ ones

320 = ______ tens ______ ones

d. 139 = ______ hundreds ______ tens ______ ones

139 = ______ tens ______ ones

e. 473 = ______ hundreds ______ tens ______ ones

473 = ______ tens ______ ones

f. 680 = ______ hundreds ______ tens

680 = ______ tens

g. 817 = ______ hundreds ______ ones

817 = ______ tens ______ ones

h. 921 = ______ hundreds ________ ones

921 = ______ tens ______ ones

2. Write down how you can skip-count by ten from 350 to 240. You might use place value disks, number lines, bundles, or numbers.

Students follow the steps of the Read, Draw, Write (RDW) process to solve word problems: Read the problem; draw and label a model of the information given; write an equation to solve; write a statement of the answer to the question.

Pencils come in boxes of 10.

a. How many boxes should Kadyn buy if he needs 136 pencils?

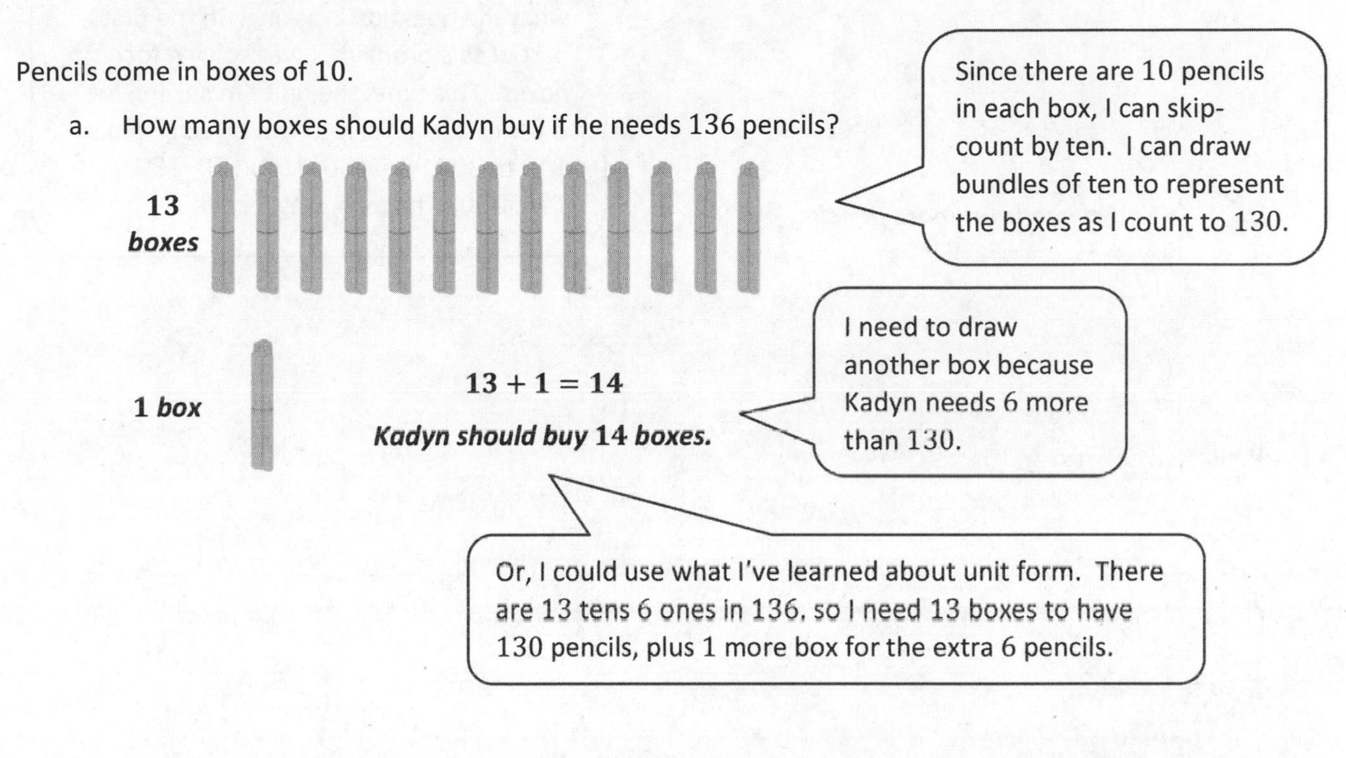

$13 + 1 = 14$

Kadyn should buy 14 boxes.

b. How many pencils will Kadyn have left over after he gets what he needs out of the boxes?

$10 - 6 = 4$

Kadyn will have 4 pencils left over.

Kadyn will use all 130 pencils from the first 13 boxes. Then, he'll need to take 6 pencils out of the last box of ten. That means 4 pencils will be left over.

c. How many more pencils does he need to have 200?

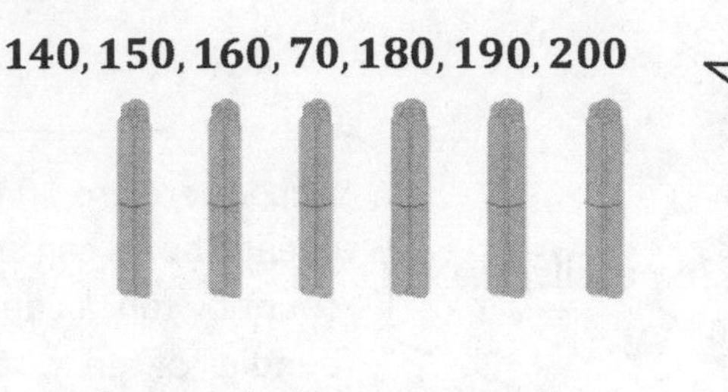

Kadyn needs 60 more pencils.

I have to be careful and pay attention to what the question is asking. In the first part of this problem, I was solving for *boxes*. This time, the unit I'm solving for is *pencils*! I can skip-count by ten from 140 to 200. So, 150, 160, 170, 180, 190, 200. That is 6 tens, or 60.

Name ______________________________ Date ______________

Pencils come in boxes of 10.

1. How many boxes should Erika buy if she needs 127 pencils?

2. How many pencils will Erika have left over after she gets what she needs out of the boxes?

3. How many more pencils does she need to have 200 pencils?

Drawing the numbers with disks on the place value chart makes it easy to compare them.

1. Draw the following numbers using place value disks on the place value charts. Answer the questions below.

   a. 132 b. 312 c. 213

(100)	(10) (10) (10)	(1) (1)

(100) (100) (100)	(10)	(1) (1)

(100) (100)	(10)	(1) (1) (1)

   d. Order the numbers from least to greatest: **132**, **213**, **312**

Hundreds are the biggest unit here, and 312 has more hundreds than the other numbers. 132 is the smallest number because it only has 1 hundred.

You could also compare all the tens in each number. 132 has 13 tens, 213 has 21 tens, and 312 has 31 tens.

2. Circle *less than* or *greater than*. Whisper the complete sentence.

a.	$300 + 60 + 5$ is (less than) / greater than 635.
b.	4 tens and 2 ones is less than / (greater than) 24.

$300 + 60 + 5 = 365$. 365 is less than 635 because it only has 3 hundreds. 635 has 6 hundreds. I could also think of it as 36 tens is less than 63 tens.

In this problem, tens are the greatest unit. 4 tens and 2 ones equals 42. 42 is greater than 24 because it has 4 tens, and 24 only has 2 tens. I could also think of it as 40 is greater than 20.

3. Write >, <, or =. Whisper the complete number sentences as you work.

a. 419 (<) 491

Place value helps me compare the numbers, especially when the digits are all the same. Both numbers have 4 hundreds, so I'm careful to notice which digit is in the tens place. 1 ten is less than 9 tens, so 419 is less than 491.

b. 908 (<) nine hundred eighty

980

When the problems are written in word form or unit form, I just rewrite them in standard form. Then, it's easy to see the digits in their places. 908 is less than 980. The hundreds are the same, but 0 tens is less than 8 tens.

c. 4 tens 2 ones (=) $30 + 12$

42

4 tens 2 ones equals 42, and $30 + 12 = 42$. That's easy! 42 equals 42.

d. $36 - 10$ (>) 2 tens 5 ones

25

$36 - 10 = 26$. 2 tens 5 ones equals 25. 26 is greater than 25.

Name ______________________________ Date ______________

1. Draw the following numbers using place value disks on the place value charts. Answer the questions below.

   a. 241 b. 412 c. 124

   d. Order the numbers from least to greatest: ________, ________, ________

2. Circle *less than* or *greater than*. Whisper the complete sentence.

a. 112 is less than / greater than 135.	d. 475 is less than / greater than 457.
b. 152 is less than / greater than 157.	e. 300 + 60 + 5 is less than / greater than 635.
c. 214 is less than / greater than 204.	f. 4 tens and 2 ones is less than / greater than 24.

3. Write >, <, or =.

   a. 100 ◯ 99

   b. 316 ◯ 361

   c. 523 ◯ 525

   d. 602 ◯ six hundred two

   e. 150 ◯ 90 + 50

   f. 9 tens 6 ones ◯ 92

   g. 6 tens 8 ones ◯ 50 + 18

   h. 84 - 10 ◯ 7 tens 5 ones

I have to read carefully! In Part (a), the ones are first, and the tens come after, but when placed on the place value chart, the hundreds come first.

When I whisper count as I draw, I see that I am comparing 112 and 115. 112 is less than 115.

1. Whisper count as you show the numbers with place value disks. Circle $>, <,$ or $=$.

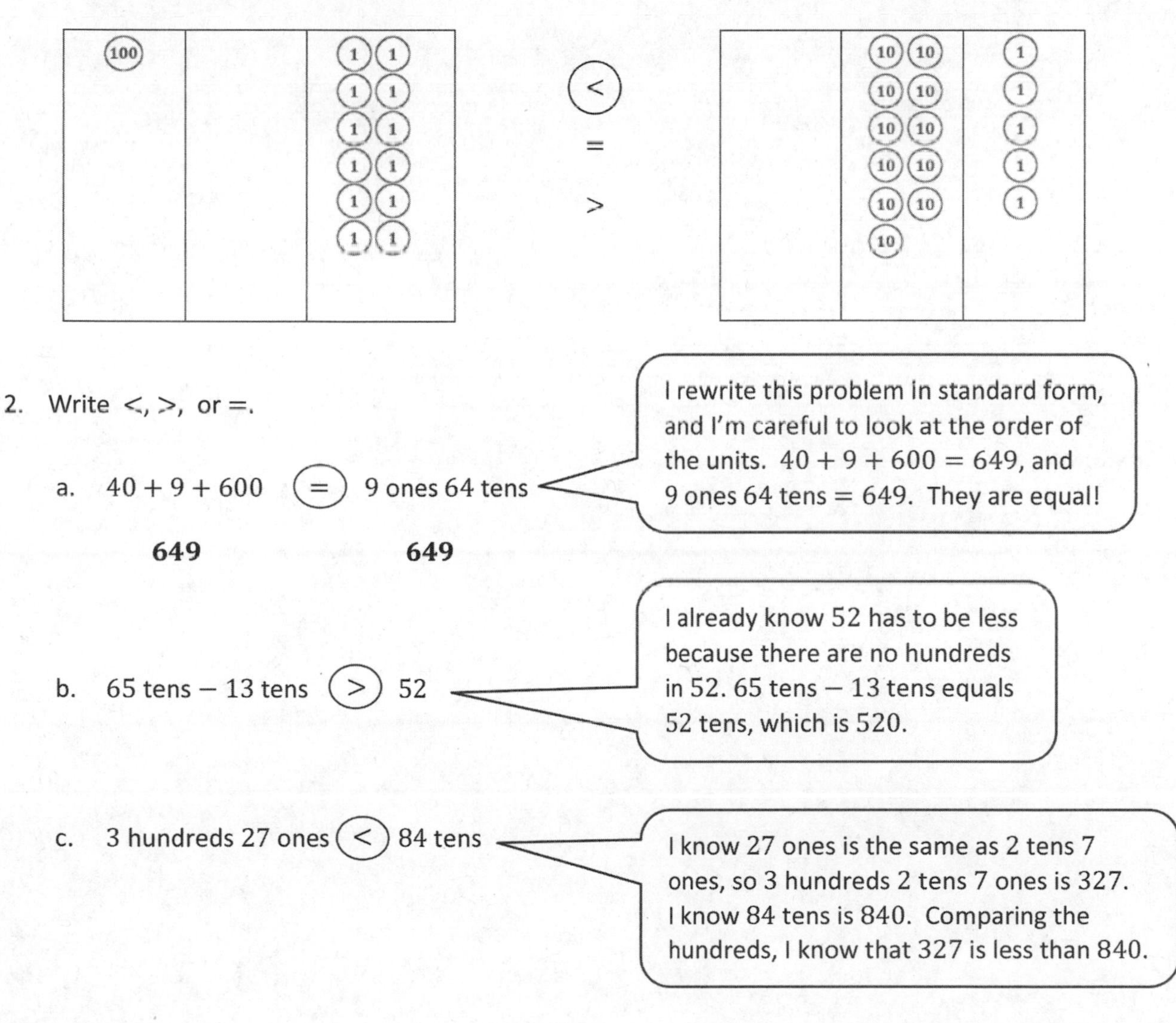

Name ______________________________ Date ______________

1. Whisper count as you show the numbers with place value disks. Circle >, <, or =.

a. Draw 13 ones and 2 hundreds.

b. Draw 12 tens and 8 ones.

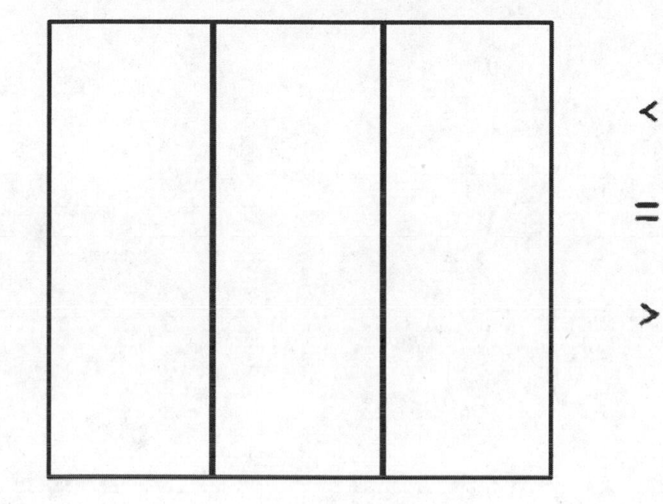

<

=

>

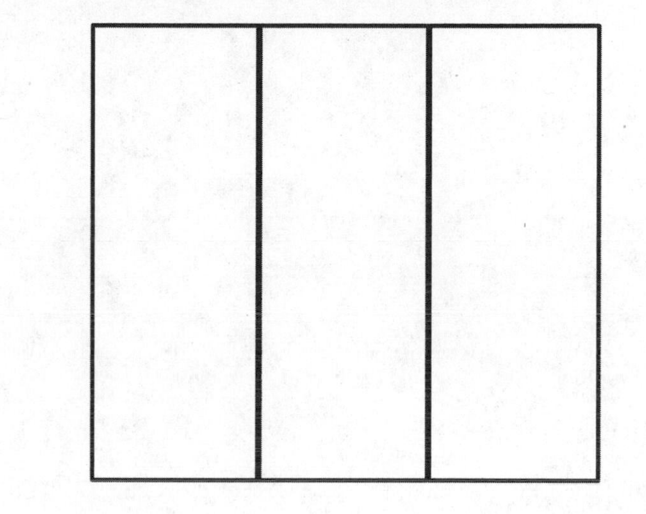

2. Write >, <, or =.

a. 199 ◯ 10 tens

b. 236 ◯ 23 tens 5 ones

c. 21 tens ◯ Two hundred twenty

d. 380 ◯ 3 hundred 8 tens

e. 20 + 4 + 500 ◯ 2 ones 45 tens

f. 600 + 7 ◯ 76 tens

g. 400 + 2 + 50 ◯ 524

h. 59 tens + 2 tens ◯ 610

i. 506 ◯ 50 tens

j. 97 tens - 12 tens ◯ 85

k. 67 tens + 10 tens ◯ 7 hundreds 7 ones

l. 8 hundreds 13 ones ◯ 75 tens

> I could draw these numbers in many different ways, but I want to be efficient. Drawing this way also makes it really easy to compare the numbers.

1. Draw the following values on the place value charts as you think best.

a. 123

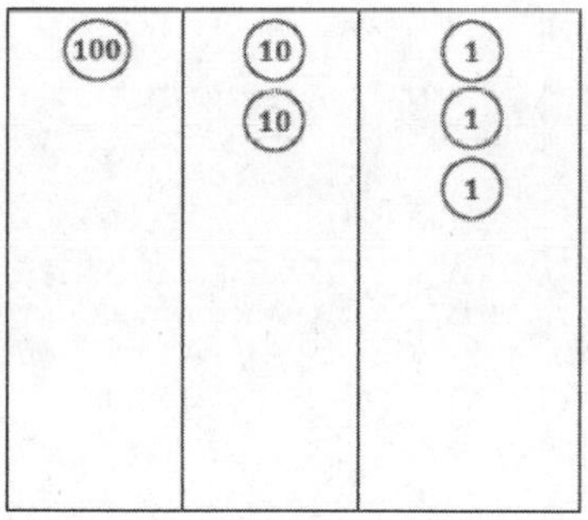

(100)	(10) (10)	(1) (1) (1)

b. 321

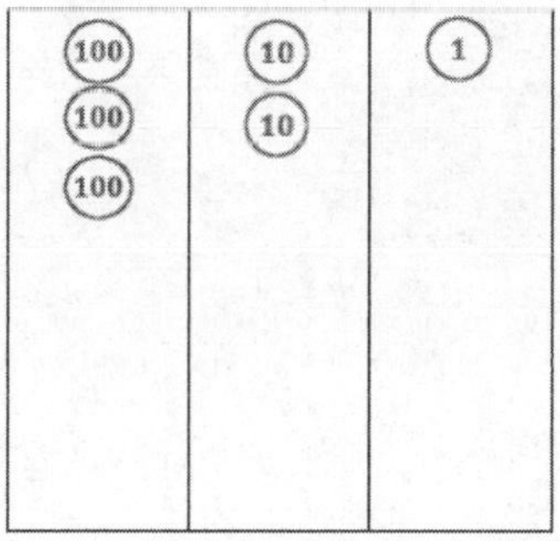

(100) (100) (100)	(10) (10)	(1)

c. 231

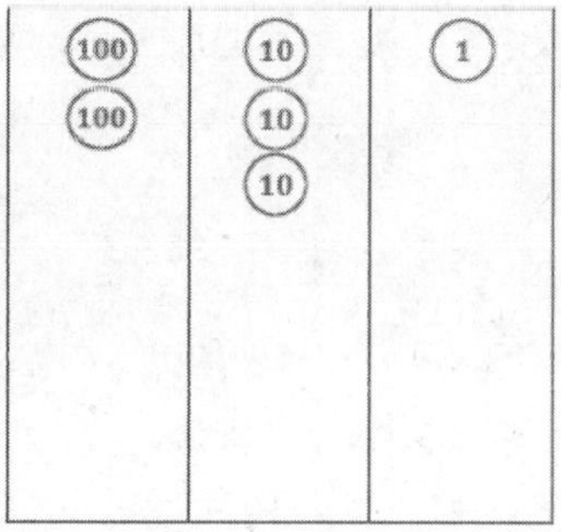

(100) (100)	(10) (10) (10)	(1)

d. Order the numbers from least to greatest: **123** , **231** , **231**

> I can see that 123 has the fewest hundreds, so It Is the smallest number. 321 has the most hundreds, so that means it's the biggest number. And 231 is in between.

2. Order the following from least to greatest in standard form.

three hundred seventy 317 30 tens 7 ones **307**, **317**, **370**

370 **307**

> Writing the numbers in standard form helps me see the value. I see that I am comparing 370, 317, and 307.

> Since the hundreds are the same, I compare the tens.

3. Order the following from greatest to least in standard form.

> Careful! This time, the order is from greatest to least.

4 ones 6 hundreds 46 tens + 10 tens 640 **640**, **604**, **560**

604 **56** ***tens***

Name ______________________________ Date ______________

1. Draw the following values on the place value charts as you think best.

   a. 241

   b. 412

   c. 124

   d. Order the numbers from least to greatest: ________, ________, ________

2. Order the following from least to greatest in standard form.

   a. 537 263 912 ________, ________, ________

   b. two hundred thirty 213 20 tens 3 ones ________, ________, ________

   c. 400 + 80 + 5 4 + 800 + 50 845 ________, ________, ________

3. Order the following from greatest to least in standard form.

   a. 11 ones 3 hundreds 311 10 + 1 + 300 ________, ________, ________

   b. 7 ones 9 hundred 79 tens + 10 tens 970 ________, ________, ________

   c. 15 ones 4 hundreds 154 50 + 1 + 400 ________, ________, ________

1. Fill in the chart. Whisper the complete sentence: "____ more/less than ____ is ____."

> I can whisper the complete number sentence as I complete the chart.
> 100 more than 242 is 342.
> 100 less than 242 is 142.
> 10 more than 242 is 252.
> 10 less than 242 is 232.
> 1 more than 242 is 243.
> 1 less than 242 is 241.

	242	153
100 more	**342**	**253**
100 less	**142**	**53**
10 more	**252**	**163**
10 less	**232**	**143**
1 more	**243**	**154**
1 less	**241**	**152**

2. Fill in the blanks. Whisper the complete sentence.

a. 1 more than 456 is **457**.

> 1 more than 6 is 7, so 1 more than 456 is 457.

b. **100** more than 180 is 280.

> The hundreds place is now 100 more.

c. 10 less than **635** is 625.

> 10 less than what number is 625? The number I am looking for is 10 more than 625, so it must be 635.

Name ______________________________ Date ________________

1. Fill in the chart. Whisper the complete sentence: "____more/less than ___ is ___."

	146	235	357	481	672	814
100 more						
100 less						
10 more						
10 less						
1 more						
1 less						

2. Fill in the blanks. Whisper the complete sentence.

a. 1 more than 103 is __________.

b. 10 more than 378 is __________.

c. 100 less than 545 is ________.

d. ________ more than 123 is 223.

e. ________ less than 987 is 977.

f. ________ less than 422 is 421.

g. 1 more than __________ is 619.

h. 10 less than __________ is 546.

i. 100 less than _________ is 818.

j. 10 more than _________ is 974.

1. Fill in the blanks. Whisper the complete sentence.

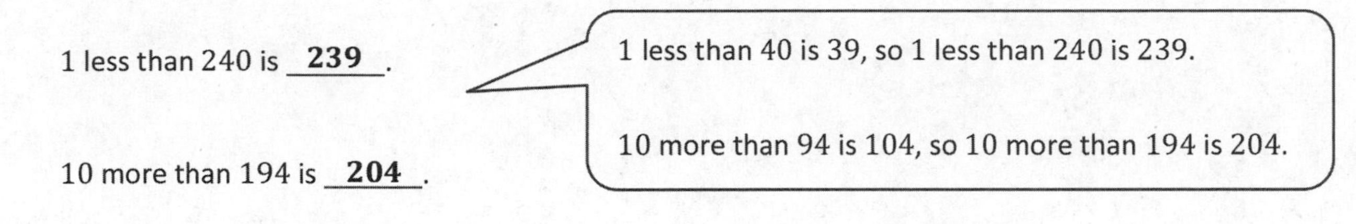

1 less than 240 is __239__.

10 more than 194 is __204__.

> 1 less than 40 is 39, so 1 less than 240 is 239.
>
> 10 more than 94 is 104, so 10 more than 194 is 204.

> I can look to see what changed. 239 changed to 240. 240 is 1 more than 239.
>
> 497 changed to 507. 507 is 10 more than 497.

__1__ more than 239 is 240.

__10__ more than 497 is 507.

10 more than __292__ is 302.

> I can think 10 more than what number is 302? So the number I am looking for is 10 less than 302. That's 292.

2. Whisper the numbers as you count.

> I can count by 1's, 10's. and 100's.

a. Count by 1's from 396 to 402. **396, 397, 398, 399, 400, 401, 402**

b. Count by 10's from 396 to 456. **396, 406, 416, 426, 436, 446, 456**

c. Count by 100's from 396 to 996. **396, 496, 596, 696, 796, 896, 996**

Name ___________________________ Date ______________

1. Fill in the blanks. Whisper the complete sentence.

   a. 1 less than 160 is ________.

   b. 10 more than 392 is ________.

   c. 100 less than 425 is ________.

   d. ________ more than 549 is 550.

   e. ________ more than 691 is 701.

   f. 10 more than ________ is 704.

   g. 100 less than ________ is 986.

   h. 10 less than ________ is 815.

2. Count the numbers aloud to a parent:

   a. Count by 1s from 204 to 212.

   b. Skip-count by 10s from 376 to 436.

   c. Skip-count by 10s from 582 to 632.

   d. Skip-count by 100s from 908 to 8.

3. Henry enjoys watching his pet frog hop.

   Each time his frog hops, Henry skip-counts backward by 100s.

   Henry starts his first count at 815.

   How many times dose his frog have to jump to get to 15?

   Explain your thinking below.

1. Find the pattern. Fill in the blanks.

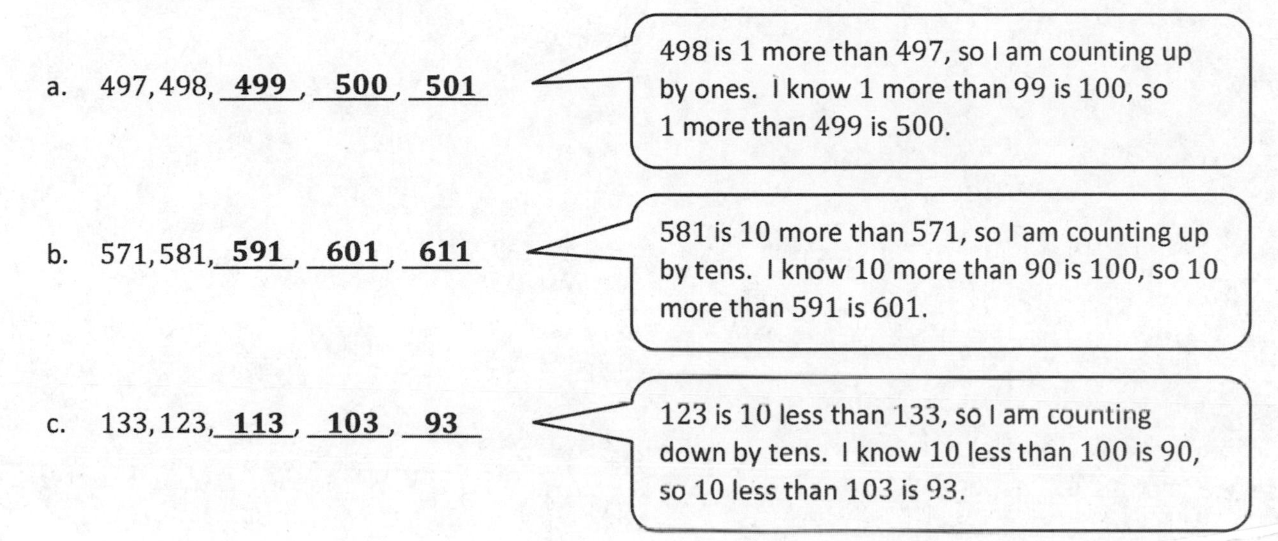

a. 497, 498, **499**, **500**, **501**

b. 571, 581, **591**, **601**, **611**

c. 133, 123, **113**, **103**, **93**

2. Fill in the chart.

I can count 1 more or 1 less as I move across the chart. 1 more than 345 is 346. 1 less than 366 is 365. Once I know the pattern, it is easy to complete the chart.

This puzzle has a pattern! It is like a hundreds chart. I can count 10 more when I move down the chart. 10 more than 348 is 358.

	345 →	346	**347**	348 ↓	**349**
		356	**357**	**358**	
364	**365**	← 366	**367**	368	**369**
		376	377		

Name ________________________________ Date ________________

1. Find the pattern. Fill in the blanks.

   a. 396, 397, __________, __________, __________, __________

   b. 251, 351, __________, __________, __________, __________

   c. 476, 486, __________, __________, __________, __________

   d. 630, 620, __________, __________, __________, __________

   e. 208, 209, __________, __________, __________, 213

   f. 316, __________, __________, 616, 716, __________

   g. 547, __________, 527, __________, 507, __________

   h. 672, __________, 692, __________, __________

2. Fill in the chart.

	206				
			218		
					230
		237			

Credits

Great Minds® has made every effort to obtain permission for the reprinting of all copyrighted material. If any owner of copyrighted material is not acknowledged herein, please contact Great Minds for proper acknowledgment in all future editions and reprints of this module.

- Page 47 (bottom), © ciroorabona/Fotolia.com